# Fundamentals of Optical Parametric Processes and Oscillators

# Laser Science and Technology
## An International Handbook

---

**Editors in Chief**



---

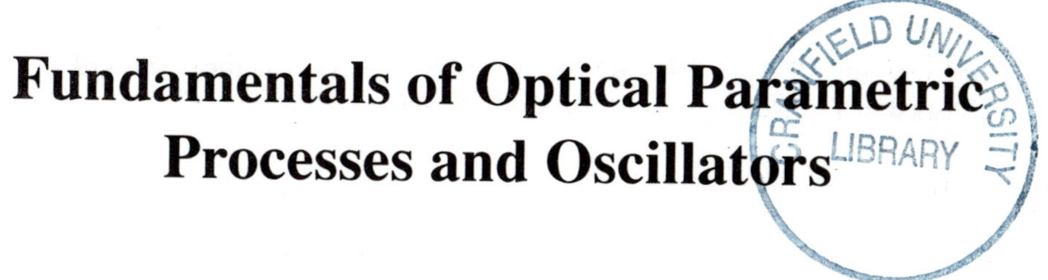

# Fundamentals of Optical Parametric Processes and Oscillators



C.L. Tang

*School of Electrical Engineering, Cornell University, Ithaca, USA*

and

L.K. Cheng

*Science and Engineering Laboratory, DuPont Company, Wilmington, Delaware, USA*

**harwood academic publishers**

Australia China France Germany India Japan
Luxembourg Malaysia Russia Singapore Switzerland
Thailand The Netherlands United Kingdom United States

Published in The Netherlands by Harwood Academic Publishers GmbH.

 Printed in Malaysia.

Emmaplein 5
1075 AW Amsterdam
The Netherlands

---

**British Library Cataloguing in Publication Data**

Tang, C. L.
Fundamentals of Optical Parametric
Processes and Oscillators. — (Laser
Science Technology Series,
ISSN 0899–2711; Vol. 10)
I. Title II. Cheng, L. K. III. Series
621.36

ISBN 3–7186–5818–6

# CONTENTS

**Introduction to the Series** vii

**1. Introduction** 1

**2. Classical Theory of Optical Parametric Amplification** 9
- 2.1 Phase-matching conditions and tuning characteristics 9
- 2.2 Parametric amplification as a repeated difference-frequency generation process 15
- 2.3 Coupled-wave and coupled-amplitude equations 17
- 2.4 Large-signal theory of parametric interaction 23
- 2.5 Bandwidth of the optical parametric process 25

**3. Quantum Theory of Optical Parametric Processes** 29
- 3.1 The field Hamiltonian for the optical parametric process 30
- 3.2 Transition probability and the initial and final states 32
- 3.3 Spontaneous parametric emission 33

**4. Optical Parametric Oscillators** 37
- 4.1 Oscillation threshold conditions 38
- 4.2 Large signal theory of parametric oscillators 42
- 4.3 Practical efficiency considerations 46
- 4.4 Parametric oscillator tuning and linewidth considerations 49
- 4.5 Optical damage 51

**5. Materials, Properties and Characterization** 53
- 5.1 Material requirements 53
- 5.2 Material survey 67
- 5.3 Mid infrared OPO materials 68
- 5.4 Visible and near-infrared OPO materials 70
- 5.5 Visible and UV OPO materials 85
- 5.6 Additional remarks 94

**6. Examples of Practical Optical Parametric Oscillators and Amplifiers** 101
6.1 High-energy nanosecond parametric oscillators 102
6.2 Continuous-pulse-train, high repetition rate, femtosecond parametric oscillators 106
6.3 Picosecond optical parametric oscillators and amplifiers 123

**Index** 129

# Introduction to the Series

Almost 30 years have passed since the laser was invented; nevertheless, the fields of lasers and laser applications are far from being exhausted. On the contrary, during the last few years they have been developing faster than ever. In particular, various laser systems have reached a state of maturity such that more and more applications are seen suffusing fields of science and technology, ranging from fundamental physics to materials processing and medicine. The rapid development and large variety of these applications call for quick and concise information on the latest achievements; this is especially important for the rapidly growing inter-disciplinary areas.

The aim of *Laser Science and Technology – An International Handbook* is to provide information quickly on current as well as promising developments in lasers. It consists of a series of self-contained tracts and handbooks pertinent to laser science and technology. Each tract starts with a basic introduction and goes as far as the most advanced results. Each should be useful to researchers looking for concise information about a particular endeavor, to engineers who would like to understand the basic facts of the laser applications in their respective occupations, and finally to graduate students seeking an introduction into the field they are preparing to engage in.

When a sufficient number of tracts devoted to a specific field have been published, authors will update and cross-reference their pages for publication as a volume of the handbook.

All the authors and section editors are outstanding scientists who have done pioneering work in their particular field.

*V.S. Letokhov*
*C.V. Shank*
*Y.R. Shen*
*H. Walther*

# 1. INTRODUCTION

Lasers are basically discrete-wavelength sources of coherent radiation involving stimulated emission between quantized energy levels in an amplifying medium. Only when these quantized energy levels are tunable or there are neighboring energy levels that are sufficiently broadened to merge into one another to form a continuous band can continuously tunable laser action be achieved. Even then, the tuning range tends to be limited. Ever since the invention of the laser, there has been a great deal of interest in the development of truly continuously and broadly tunable coherent radiation sources. Such sources would have broad applications in research and industry.

Optical parametric devices (OPD) are powerful solid state sources of broadly tunable coherent radiation capable of covering the entire spectral range from the near uv to the mid ir and can operate from continuous-wave down to the femtosecond time domain. They are nonlinear optical devices that convert laser radiation at one frequency to a continuously tunable spectral range through the three-photon parametric process. As a result of recent advances in the development of lasers that can serve as pump sources for OPD's and in nonlinear optical materials research, optical parametric oscillators and amplifiers are now becoming practical devices. Looking toward the future, it is quite possible that, as even better nonlinear crystals become more widely available, one of the principal applications of high-power lasers will be as the pump source for optical parametric oscillators and amplifiers, and most lasers with high enough output powers will have OPD's as an attachment to extent the wavelength range.

The history of OPD is a long one, almost as long as that of the laser.[1–3] For a variety of reasons, however, the OPD technology never quite developed to the same level of maturity as the laser technology. One obvious reason for the slow development of OPD technology is that the devices require laser pump sources. The development of suitable pump lasers clearly must precede the development of the OPD's. The second main reason is that the development of nonlinear optical crystals suitable for OPD applications is a slow and tedious process.

As a result of the steady, though slow, progress in nonlinear optical materials research over the years, the situation has finally changed culminating in the recent rapid advances in the development of a wide variety of practical optical parametric oscillators and amplifiers from continuous wave to the nanosecond, picosecond, and down to the femtosecond time domain.

The basic principles of the parametric process had been known long before the advent of the lasers and date back to the days of the masers (microwave amplification of stimulated emission of radiation). In those days, the interest was in low noise microwave amplifiers. The fundamental noise limit of the parametric amplifier[4] like that of the maser, is due to the quantum mechanical uncertainty-principle and is related to the zero-point fluctuations in the amplitude and phase channels of the amplifier. In the microwave range, the equivalent temperature of the zero-point fluctuations (hv per mode per volume), $T_{eq} = h\nu/k(\ln 2)$, is extremely low. There was, therefore, considerable interest in the microwave parametric amplifier as a low-noise amplitude- and phase-sensitive amplifier in the 1950's. In the optical region, because the noise due to zero-point fluctuations far exceeds that due to thermal fluctuations ($kT$ per mode per volume) even at room temperature, the optical parametric amplifier has a relatively high equivalent noise temperature. It is, therefore, not particularly useful as a low-noise amplitude and phase-sensitive amplifier even though it is quantum-noise limited. Based upon the recently developed concept of "squeezing", through which the quantum noise can be selectively concentrated more in either the amplitude or in the phase channel, the optical parametric amplifier may conceivably be made into a low noise amplifier by putting the signal in the alternative quiet channel, but the practicality of such a scheme is far from having been demonstrated. The real interest in the optical domain is in the wavelength tunability (Figure 1.1) of the optical parametric oscillator and amplifier as sources of coherent radiation.

The parametric process is one of the most elementary nonlinear optical processes. It involves three photons and can be represented schematically by the simplest kind of

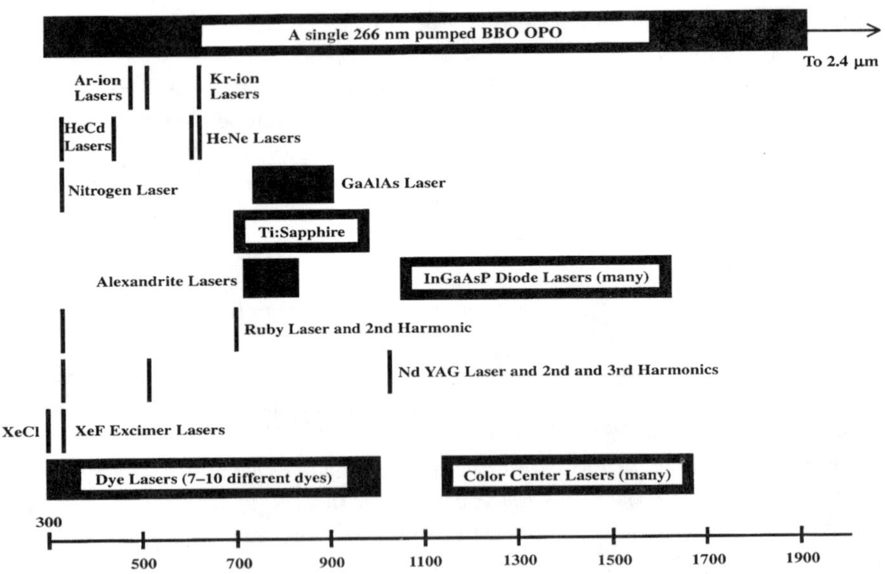


**Fig. 1.1** Example of the tuning range of an OPO compared with some other types of lasers.

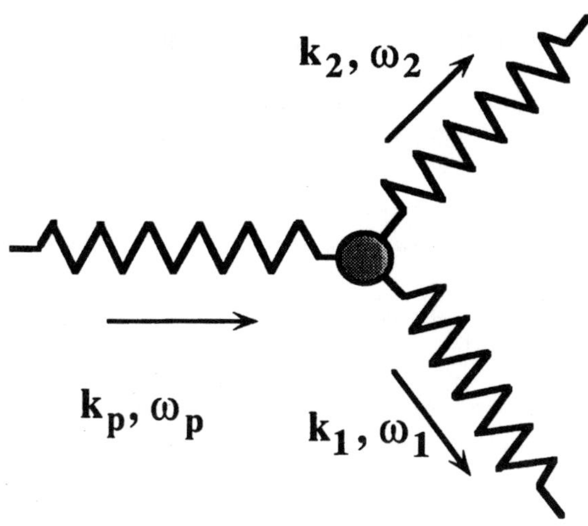


**Fig. 1.2** Feynman diagram of three-photon optical parametric process.

Feynman diagram (Figure 1.2) that describes the elementary three-photon process in which one high-frequency photon is annihilated and two lower-frequency photons are created. (For more details, see Chapter 3.) This process has its origin in the second order nonlinear polarization in the expansion of the induced macroscopic polarization **P** in the nonlinear medium in powers of the electric field **E** [cgs electrostatic units]:

$$\begin{aligned} \mathbf{P} &= \chi^{(1)} \cdot \mathbf{E} + \chi^{(2)} : \mathbf{EE} + \chi^{(3)} \vdots\, \mathbf{EEE} + \ldots.. \\ &= \mathbf{P}^{(1)} + \mathbf{P}^{(2)} + \mathbf{P}^{(3)} + \ldots\ldots \end{aligned} \tag{1.1}$$

This second order polarization term, $\mathbf{P}^{(2)}$, leads to a term proportional to $\mathbf{E}^3$ in the Hamiltonian of the field in the medium. Quantizing the field leads in turn to a term of the form $\chi^{(2)}\mathbf{a}_1^-\mathbf{a}_2^+\mathbf{a}_3^+$, where $\chi^{(2)}$ is the second order nonlinear susceptibility of the optical medium and the $\mathbf{a}^+$'s and $\mathbf{a}^-$'s are the creation and annihilation operators of the appropriate photons, respectively. A term of this form implies the possibility of creating two photons of lower frequencies from one of higher frequency. This is the three-photon parametric process.

It is obvious that the basic three-photon parametric process can be generalized to those involving four or more photons corresponding to the higher order terms in the induced polarization in the nonlinear medium. Four-photon parametric processes, also known as four-wave mixing processes, are used extensively for various fundamental measurements of atomic, molecular, or macroscopic properties of materials. Practical device applications based upon these higher order processes are not yet well developed, however. **The main focus of this monograph is on the three-photon parametric process.**

In a three-photon parametric process,[5] if the initial state, $|i\rangle$, of the field contains $N_1$ pump photons in mode-1 at frequency $\omega_1$ but no photon in mode-2 and mode-3 at frequencies $\omega_2$ or $\omega_3$, respectively, the corresponding parametric process is the spontaneous parametric emission process through which one pump photon at $\omega_1$ spontaneously breaks down into two lower frequency photons at $\omega_2$ and $\omega_3$ in the final state $|f\rangle$, or:

$$|i\rangle = |N_1 00\rangle \rightarrow |f\rangle \propto \sqrt{N_1}|(N_1 - 1)11\rangle \tag{1.2}$$

In this case, the corresponding transition probability is proportional to $N_1$ or the intensity of the pump beam. If there are already photons at $\omega_2$ and $\omega_3$ (which are also referred to as signal and idler frequencies $\omega_s$ and $\omega_i$ using the nomenclature borrowed from earlier microwave parametric amplifier work), then stimulated emission takes place, or:

$$|i\rangle = |N_1 N_2 N_3\rangle \rightarrow |f\rangle \propto \sqrt{N_1(N_2 + 1)(N_3 + 1)}|(N_1 - 1)(N_2 + 1)(N_3 + 1)\rangle, \tag{1.3}$$

which shows that the transition probability to the final state contains terms that are proportional to not only the intensity of the pump beam but also those of the signal and idler beams. This means that the more the signal or idler photons present in the medium already, the more signal and idler photons will be emitted through this process. This is just like the stimulated emission process in ordinary lasers or masers. Here it corresponds to amplification of the signal and idler beams through the parametric amplification process. Note that Eq. (1.3) reduces to the spontaneous emission result, Eq. (1.2), if $N_2$ and $N_3$ are both equal to zero.

Given the parametric amplification process, an optical parametric oscillator requires only the addition of suitable optical feedback, such as a Fabry-Perot cavity. The basic oscillator configuration is, therefore, extremely simple, as shown schematically in Figure 1.3. It is completely analogous to that of conventional lasers, except that the active medium is an optically pumped nonlinear optical crystal and no unique set of discrete energy levels of the medium is directly involved in converting the pump photons into the signal and idler photons.

The tunability of the optical parametric oscillator comes about as follows. As in any multi-photon process, energy in the process is always conserved: $\omega_1 = \omega_2 + \omega_3$. However, in going from one high frequency to two lower frequencies, the split is not unique. As long as the the sum is conserved, $\omega_2$ or $\omega_3$ can each have basically any values from zero to $\omega_1$. This flexibility is, in fact, the origin of the basic tunability

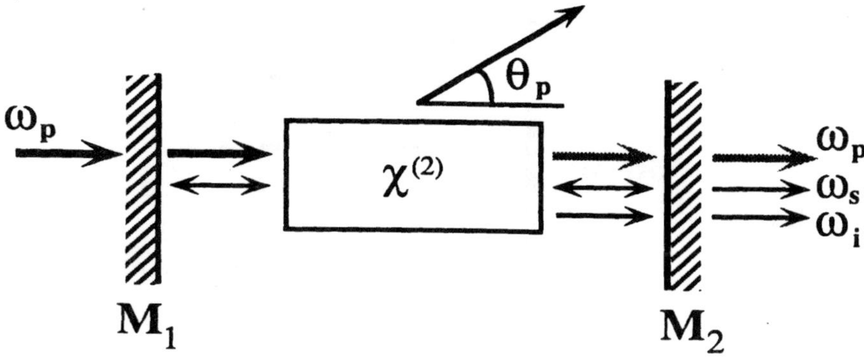


**Fig. 1.3** Schematic of singly-resonant optical parametric oscillator.

of the optical parametric process. The specific pair of frequencies resulting from each $\omega_1$ is ultimately determined by the photon momentum conservation condition also known as the phase-matching condition, $\mathbf{k}_1 = \mathbf{k}_2 + \mathbf{k}_3$, taking into account the material dispersion and the birefringence of the crystal. Tuning is achieved by rotating the crystal relative to the direction of propagation of the waves; in doing so, the corresponding birefringence can be tuned, which in turn tunes $\omega_2$ and $\omega_3$, as will be explained in more detail in Section 2.1.

The OPD is in many ways analogous to the optically-pumped 3-level laser (Figure 1.4). Thus, many of the techniques and device physics issues for lasers such as the exact mode properties, the transverse and longitudinal mode field distributions, line-narrowing schemes, noise properties, soliton formation, and effects of phase-conjugation mirrors, squeezing, and microcavity on the emission process, etc. are also relevant to the parametric devices. There are, however, also a number of important differences. Instead of discrete energy levels as in the laser, tunable photon energy levels are involved in the OPD. As shown in Figure 1.4, for the OPD's, the middle energy level can be tuned through the phase-matching condition, and the top energy level can be tuned by tuning the pump frequency. In the laser, because the radiative processes involved are single-photon transitions and because of the possible involvement of nonradiative transitions, the temporal and spatial coherence properties of the laser output are not directly related to those of the pump radiation. In the OPD, there is no non-radiative transitions involved and the whole process is a single three-photon process. The temporal and spatial coherence properties of the OPD are, therefore, directly related to those of the pump. The pump source for a laser can, therefore, be a coherent or an incoherent source, whereas the pump source of the OPD must be a laser. Also, because the laser involves three single-step resonant

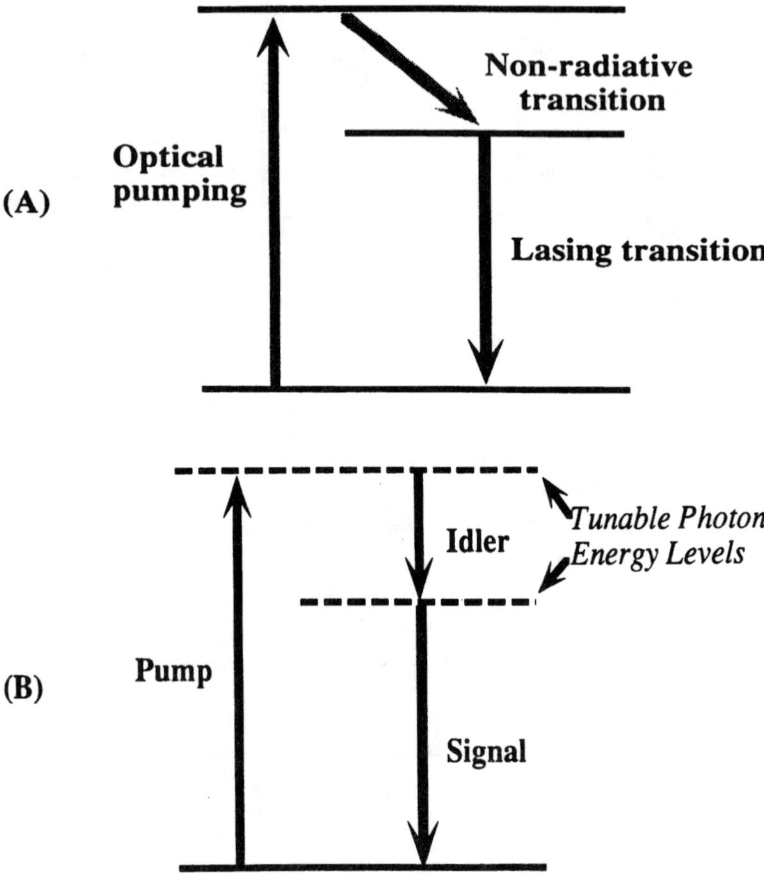


**Fig. 1.4** (A) Optically pumped three-level laser, and (B) three-photon optical parametric process.

transitions while the parametric process involves a single three-photon transition, the cross-sections for the parametric process are generally much smaller than those for the stimulated emission process in lasers. The threshold for oscillation is, therefore, generally much higher for OPO's than that for the lasers. On the other hand, above the threshold for oscillation, the OPO can be a very efficient device because there is no non-radiative process involved; there is in principle no loss. The quantum efficiency of the OPO can potentially be 100% with all the absorbed pump photons converted to useful output photons.

Turning now to a brief consideration of the nonlinear optical material requirements. The practical development of the OPD depends on the availability of suitable pump laser sources and nonlinear crystals. In the decades since the laser was first invented, many varieties of lasers have been developed. Many of these can potentially be used as pump sources for OPD's if suitable nonlinear optical crystals are available. For OPD applications, a most important property of the nonlinear crystal is of course the optical nonlinearity. It must be emphasized, however, that there are many crystals that have relatively large nonlinear optical coefficients but are nevertheless not very useful. That is because there are in fact many other material properties that are equally important, such as mechanical hardness, chemical and thermal stability, optical damage threshold, phase-matching properties, etc. (see, for example, Tables 5.1–5.6 below). For a nonlinear optical crystal to be useful, it must meet some minimum requirements in all these properties. Failure in meeting any of the basic requirements is enough to make the material irrelevant no matter how nonlinear the crystal is. The recent rapid development of the OPD technology is largely a result of the fact that such new materials as BBO and KTP can now finally meet all the basic requirements for OPD applications, but these materials are still far from ideal. Thus, with the availability of better materials expected in the future, significant improvements in the technology are sure to come.

In this monograph, we will first outline the basic theory of optical parametric processes. This will be followed by discussions of the materials and practical device considerations. The emphasis here will be on optical parametric oscillators. In a companion monograph, Professor Y.R. Shen will give a detailed discussion of the equally important subject of optical parametric amplifiers.

## REFERENCES


1. J.A. Armstrong, N. Bloembergen, J. Ducuing, and P.S. Pershan, *Phys. Rev.*, **127**, 1918 (1962); also, N. Bloembergen, *Nonlinear Optics*, New York: W. A. Benjamin, Inc., 1965.
2. For references to early American works, see, for example, R.L. Byer, "Optical parametric oscillators", in *Treatise in Quantum Electronics*, volume 1, *Nonlinear Optics*, Pts. A and B, H. Rabin and C.L. Tang, Eds., New York: Academic Press, 1975. For references to early Russian works, see, for example, S.A. Akhmanov and R.V. Khokholov, *Problems of Nonlinear Optics*, New York: Gordon and Breach, 1972.
3. See, for example, Y.R. Shen, *Principles of Nonlinear Optics*, New York: Wiley Interscience, 1984, and references therein.
4. See, for example, W.H. Louisell, *Coupled Mode and Parametric Electronics*, New York: Wiley, 1960.
5. See, for example, C.L. Tang, "*Spontaneous and stimulated parametric processes*", in *Treatise in Quantum Electronics*, Vol. 1, *Nonlinear Optics*, Pts. A and B, H. Rabin and C.L. Tang, Eds., New York: Academic Press, 1975.

# 2. CLASSICAL THEORY OF OPTICAL PARAMETRIC AMPLIFICATION

Even though the basic spontaneous parametric process is a purely quantized-field effect,[1] the optical parametric amplification process, or the stimulated parametric process, can be understood within the context of classical electromagnetic theory without invoking any quantum field theory concepts. The process can also be viewed qualitatively as a repeated difference-frequency process. Unlike in the sum-frequency process, in which the generation of a photon at the sum-frequency consumes input photons at both input frequencies, in the difference-frequency process, for each difference-frequency photon generated at $\omega_3$ from an $\omega_1$ photon and an $\omega_2$ photon, an additional lower frequency photon at $\omega_2$ is also generated and added to the incident beam at $\omega_2$. Repeating this difference-frequency process as the three waves continue to propagate in the medium, the more $\omega_2$ photons are generated, the more $\omega_3$ photons are generated; the more $\omega_3$ photons are generated, still more $\omega_2$ photons are generated, leading to amplification of the incident wave at $\omega_2$ at the expense of the "pump beam" at $\omega_1$. Using the terminology originally adopted for microwave parametric amplifier work, the wave being amplified is usually referred to as the "signal wave", while the difference-frequency wave at $\omega_3$ that is also generated as a by-product in the parametric amplification process is sometimes referred to as the "idler" wave. In the difference-frequency generation process, the photons generated at $\omega_3$ are the end product of interest; the term "idler wave" would appear to be a misnomer in this case. The repeated-difference-frequency generation process can be used as a model for analyzing the parametric amplification process.

## 2.1 PHASE-MATCHING CONDITION AND TUNING CHARACTERISTICS

As in any nonlinear optical process, the energy and momentum of the interacting photons must be conserved. In the three-photon parametric process, the conditions are:

$$\omega_p = \omega_s + \omega_i \tag{2.1}$$

$$\mathbf{k_p} = \mathbf{k_s} + \mathbf{k_i}. \tag{2.2}$$

The basic tuning characteristics of the parametric process is completely determined by these conditions. As pointed out before, the tunability of the optical parametric process comes from the fact that, when one fixed high-frequency photon breaks down and converts into two lower-frequency photons, the resulting frequencies satisfying the energy-conservation condition, Eq. (2.1), are not unique. The specific pair of signal and idler frequencies that will be produced is determined by the momentum-conservation, or phase-matching, condition Eq. (2.2) in the following way. Since $k = n\omega/c$, if the medium is nondispersive and isotropic, once the energy conservation condition is satisfied, the phase-matching or momentum-conservation condition is automatically satisfied for collinear interaction:

$$k_p = k_s + k_i, \tag{2.3}$$

where $k$ is the magnitude of wave vector $\mathbf{k}$. However, all optical materials are dispersive. In general, the energy and momentum conservation conditions cannot be satisfied simultaneously due to material dispersion. This is particularly true in the case of parametric processes where the pump, signal, and idler frequencies can be very different from each other. In the normally dispersive region, the index-of-refraction generally increases with frequency; thus, the magnitude of the $\mathbf{k}$-vector of the pump wave is generally too long (Figure 2.1). Some special scheme must be used to reduce $k_p$ in order to satisfy the phase-matching condition. The commonly used method is to use the birefringence in anisotropic crystals to compensate for the material dispersion.

Consider, for example, a negative uniaxial crystal. As shown in Figure 2.2, to make $k_p$ equal to $(k_s + k_i)$, one can make the pump wave an extraordinary wave and the signal and idler waves ordinary waves. If the birefringence is larger than the dispersion, in the sense $|n^{(0)}(\omega_p) - n^{(e)}(\omega_p)| \geq |n^{(0)}(\omega_p) - n^{(o)}(\omega_s)\omega_s/\omega_p - n^{(o)}(\omega_i)\omega_i/\omega_p|$ there will be a set of specific frequencies such that $k_p^{(e)}$ is equal to $k_s^{(o)} + k_i^{(o)}$ at some angle, $\theta_p$, relative to the optic axis. These are the signal and idler frequencies for the particular pump frequency and pumping direction. The signal and idler frequencies can then be tuned simply by rotating the crystal relative to the direction of propagation of the pump wave as defined by $\mathbf{k}_p^{(e)}$. The phase-matching condition corresponding to matching a pump wave that is polarized orthogonal to both the signal and an idler waves (e.g. $k_p^{(e)} = k_s^{(o)} + k_i^{(o)}$, for negative uniaxial crystals) is called Type-I phase-matching. The corresponding phase-matching angle for uniaxial crystals in the case of collinear interaction can be determined from the following equation:

$$\frac{\cos^2\theta_p}{n_p^{(o)2}} + \frac{\sin^2\theta_p}{n_p^{(e)2}} = \frac{\omega_p^2}{\left[n_s^{(o)}\omega_s + n_i^{(o)}\omega_i\right]^2} \tag{2.4}$$

for Type-I phase-matching. Practical examples of tuning curves based on Type-I phase-matching are shown in Figure 2.3.

Another possible phase-matching condition is to have either the signal or the idler wave polarized in the same way as the pump wave (e.g. or $k_p^{(e)} = k_s^{(e)} + k_i^{(o)}$

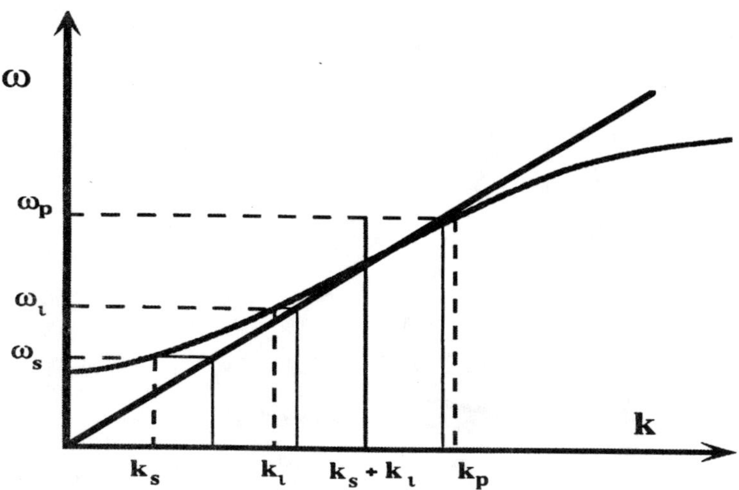


**Fig. 2.1** Due to the normal dispersion in the medium, the magnitude of the *k*-vector for the pump wave is always too long to match the sum of the *k*-vectors of signal and idler waves.

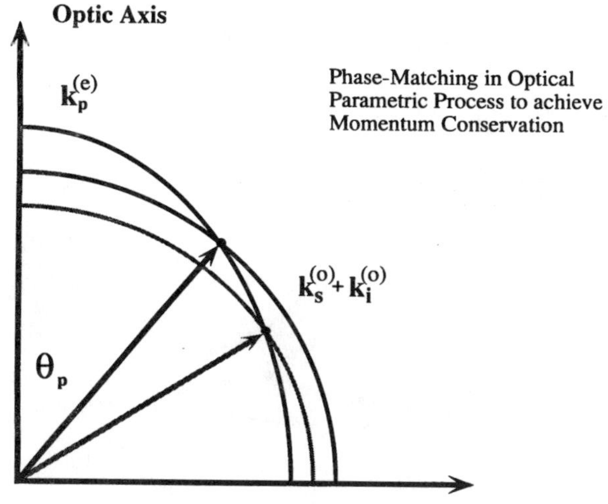


**Fig. 2.2** Use of birefringence to compensate for material dispersion. Rotating the crystal relative to the direction of propagation of the waves leads to tuning of the frequencies of the signal and idler waves.

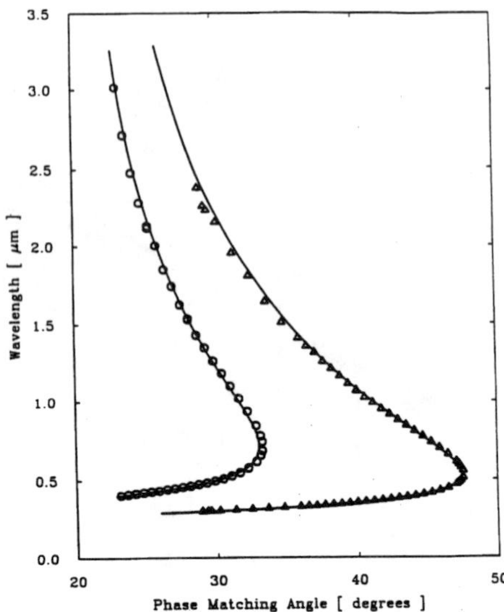


**Fig. 2.3** Type I, 355 nm pumped (O) and 266 nm (Δ) BBO tuning curves determined via the parametric luminescence techniques. The solid lines are calculated results.

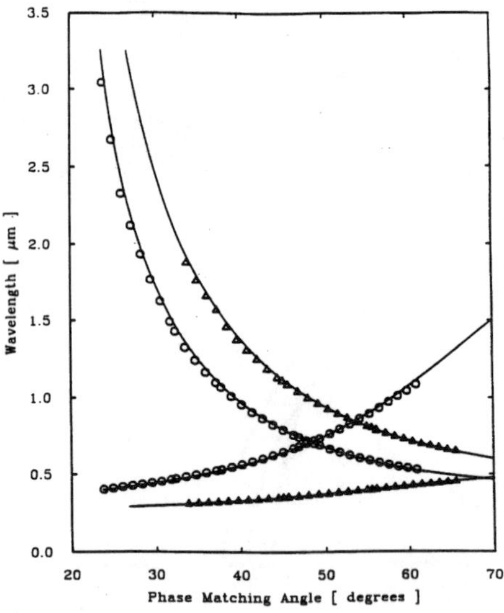


**Fig. 2.4** Type II, 355 nm pumped (O) and 266 nm pumped (Δ) BBO OPO tuning curves determined via the parametric luminescence techniques. Solid lines are corresponding calculated results.

or $k_p^{(e)} = k_s^{(o)} + k_i^{(e)}$). This type of phase-matching condition is called Type-II phase-matching condition. The corresponding phase-matching angle for the extraordinary signal wave case, for example, must be determined from the simultaneous equations:

$$\frac{\cos^2\theta_p}{n_p^{(o)2}} + \frac{\sin^2\theta_p}{n_p^{(e)2}} = \frac{\omega_p^2}{\left[n_s(\theta_p)\omega_s + n_i^{(o)}\omega_i\right]^2}, \tag{2.5}$$

$$\frac{\cos^2\theta_p}{n_s^{(o)2}} + \frac{\sin^2\theta_p}{n_s^{(e)2}} = \frac{1}{n_s(\theta_p)^2}. \tag{2.6}$$

In the case of positive uniaxial crystals, Type-I phase-matching requires that the condition $k_p^{(o)} = k_s^{(e)} + k_i^{(e)}$ be satisfied and Type-II requires that the condition $k_p^{(o)} = k_s^{(o)} + k_i^{(e)}$ or $k_p^{(o)} = k_s^{(e)} + k_i^{(o)}$ be satisfied. The phase-matching equations above must be modified accordingly. Examples of tuning curves based upon Type-II phase-matching condition are shown in Figure 2.4.

For biaxial crystals, the indices will have not only a polar angular $\theta$-dependence but also an azimuthal angular $\phi$-dependence and the phase-matching equations become correspondingly more complicated but can be obtained on the basis of the same principles.

Also, the interacting waves do not have to be collinear, as long as the vector phase-matching condition, (2.2), is satisfied, which results in the angular dependence of the spontaneous parametric emission wavelength being determined by a noncollinear phase-matching condition. Examples of the tuning characteristics of spontaneous emission corresponding to non-collinear interaction are shown in Figure 2.5.

For bulk crystals, the use of birefringence to compensate for material dispersion is most widely used to satisfy the phase-matching requirement in optical parametric processes and, in fact, most other three-photon nonlinear optical processes. Wavelength or frequency tuning is achieved through either angular tuning the birefringence or through temperature tuning the indices of the crystal. In practice, either method is simple to implement. In the case of the angular tuning scheme, it only requires a rotation mechanism for the crystal. Practical devices that are continuously tunable over a very large spectral range with computer control are now available commercially. For example, a BBO ($\beta$-barium borate) pumped by the third harmonic (355 nm) of a Nd:YAG laser can be tuned continuously from 400 nm to 2.5 $\mu$m with a single set of mirrors resonating the signal. The temperature tuning scheme has the advantage of not requiring realignment of the cavity since no mechanical change is involved in the tuning process. This is particularly important in the case of "noncritical" phase-matching in which the waves are phase-matched along a principal axis and is not sensitive to small angular deviations.

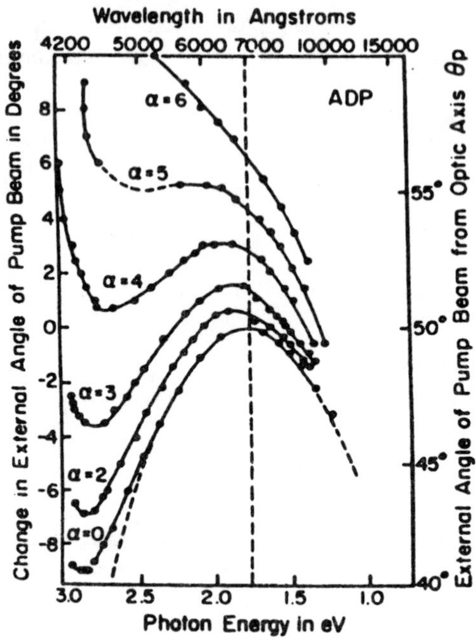


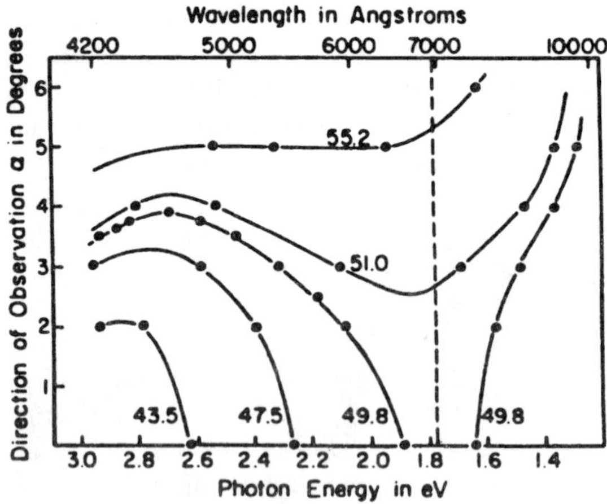


**Fig. 2.5** Examples of noncolinear tuning curve showing measured spontaneous emission in ADP ($NH_4H_2PO_4$) pumped by the second harmonic of a ruby laser (from Ref. 4).

In the case of thin-film waveguides, it is also possible to use waveguide dispersion to compensate for material dispersion. Quasi-phase-matching[2,3] making use of the dispersion of spatial harmonics of artificial periodic structures either in bulk crystals or in thin-film waveguides can also be employed to compensate for material dispersion. For practical parametric device applications, these more sophisticated phase-matching methods have so far only limited applicability, but rapid progress is being made using electric field poled $LiNbO_3$ by M. Fejer and R. Byer of Stanford University.

## 2.2 PARAMETRIC AMPLIFICATION AS A REPEATED DIFFERENCE-FREQUENCY GENERATION PROCESS

Once the phase-matching condition is satisfied, the parametric interaction process will add coherently spatially. To demonstrate the possibility of gain, the basic coupled-mode equations for the parametric amplification process will first be derived in this section on the basis of the simple picture of repeated difference-frequency generation. The more rigorous treatment of the parametric amplification process based upon Maxwell's equations taking into account the nonlinear polarization terms in the optical medium will be developed in the next section.

Consider two plane waves incident on a differential length of nonlinear optical medium from $z = 0$ to $z = \Delta z$, where $\Delta z$ is much smaller than any one of the relevant wavelengths. In the difference-frequency process, the change in the complex amplitude $\Delta E_2$(or $\Delta E_3$) of the wave at the difference-frequency, $\omega_2 = \omega_1 - \omega_3$, in the nonlinear medium with a second order nonlinear optical susceptibility $\chi^{(2)}$ is proportional to the complex amplitudes of the incident waves $E_1$ and $E_3^*$(or $E_2^*$), $\Delta z$, and the nonlinear susceptibility $\chi^{(2)}(\omega_2 = \omega_1 - \omega_3)$ [or $\chi^{(2)}(\omega_3 = \omega_1 - \omega_2)$] in the limit of low conversion or $E_3$ (or $E_2$) $\ll E_1$:

$$\Delta E_2 = \xi_2 \chi^{(2)} E_1 E_3^* \Delta z, \tag{2.7}$$

and

$$\Delta E_3 = \xi_3 \chi^{(2)} E_1 E_2^* \Delta z, \tag{2.8}$$

The proportionality constants $\xi_2$ and $\xi_3$ cannot be fixed on the basis of this kind of simple phenomenological argument. They can be determined, however, from the appropriate coupled-wave equations as will be shown later. For arbitrary values of $z$, the spatial variations of the **E**-fields must be taken into account, because as the waves propagate the spatial dependence of the complex amplitude of the **E**-fields varies with additional phase factors of the form $e^{ikz}$. If, however, the phases of the waves are matched or $\mathbf{k}_1 = \mathbf{k}_2 + \mathbf{k}_3$, the validity of Eqs. (2.7) and (2.8) can be extended and the difference-equations can be converted to differential equations that are valid for all values of $z$:

$$\frac{\partial E_2(z)}{\partial z} = \xi_2 \chi^{(2)} E_1 E_3^*(z), \tag{2.9}$$

and

$$\frac{\partial E_3(z)}{\partial z} = \xi_3 \chi^{(2)} E_1 E_2^*(z). \tag{2.10}$$

These equations are sometimes also referred to as the coupled-mode equations for the optical parametric amplification process.

Neglecting the saturation effect or the depletion of the pump-wave due to conversion to the signal and idler waves, $E_1$ can be considered to be a fixed parameter equal to the incident pump-wave amplitude at the input. Combining Eqs. (2.9) and (2.10) gives:

$$\frac{\partial^2}{\partial z^2} E_2(z) = g^2 E_2(z) \tag{2.11}$$

where

$$g = \sqrt{\xi_2 \xi_3^*} |\chi^{(2)}| \, |E_1| \tag{2.12}$$

Solving Eq. (2.11) subject to the boundary conditions that there is no reflection at the output end of the nonlinear crystal and that the input signal $E_2(0) = E_s(0)$, one finds the spatial variation of the signal wave:

$$E_2(z) = E_s(0) \cosh gz. \tag{2.13}$$

It shows that $g$ has the meaning of a spatial gain coefficient and the signal wave is indeed amplified with a cosh $gz$ spatial dependence, if the constant $\xi_2 \xi_3^*$ is real.

From Eqs. (2.10) and (2.13), and with the boundary condition that $E_3(0) = 0$ at the input, it follows that :

$$E_3^*(z) = \sqrt{\frac{\xi_3^* \chi^{(2)*} E_1^*}{\xi_2 \chi^{(2)} E_1}} E_s(0) \sinh gz. \tag{2.14}$$

This result gives the spatial variation of the idler field at the difference-frequency $\omega_3$.

In the limit of a small distance from the input end, $E_3^*(\Delta z)$ can be viewed as the field generated at the difference-frequency $\omega_3$ due to mixing of the pump field at $\omega_1$ and the input field at $\omega_2$, since the boundary condition used corresponds to no input at $\omega_3$. Taking Eq. (2.14), to the limit of small $gz$, or $gz \rightarrow g\Delta z$, it shows that the intensity $I_3(\Delta z)$ indeed has the form corresponding to difference-frequency generation from $I_1$ and $I_2(0)$:

$$I_3(\Delta z) \propto |\chi^{(2)}|^2 \Delta z^2 I_1 I_2(0). \tag{2.15}$$

This result is, therefore, consistent with the picture of viewing the parametric process as a repeated difference-frequency process.

## 2.3 COUPLED-WAVE AND COUPLED-AMPLITUDE EQUATIONS

A more complete derivation of the results given above would have to come from Maxwell's equations which in the context of the present problem reduce to the basic wave equation for the **E**-field in the nonlinear optical medium (esu system):

$$\nabla^2 \mathbf{E}(\mathbf{r}, t) - \mu_0 \varepsilon_0 \frac{\partial^2}{\partial t^2} \mathbf{E}(\mathbf{r}, t) = \frac{4\pi}{c^2} \frac{\partial^2}{\partial t^2} \mathbf{P}(\mathbf{r}, t). \tag{2.16}$$

In the case of the basic three-photon parametric process, the **E**-field in the medium contains three spectral components:

$$\begin{aligned} \mathbf{E}(\mathbf{r}, t) = {} & \frac{1}{2}\{\mathbf{E}_1(\mathbf{r}) \exp[i\mathbf{k}_1 \cdot \mathbf{r} - i\omega_1 t] + \mathbf{E}_2(\mathbf{r}) \exp[i\mathbf{k}_2 \cdot \mathbf{r} - i\omega_2 t] \\ & + \mathbf{E}_3(\mathbf{r}) \exp[i\mathbf{k}_3 \cdot \mathbf{r} - i\omega_3 t] + \text{complex conjugates}\}. \end{aligned} \tag{2.17}$$

The corresponding induced macroscopic polarization in the medium consists of a linear and a nonlinear term:

$$\mathbf{P}(\mathbf{r}, t) = \frac{1}{2}\{\mathbf{P}^L(\mathbf{r}, t) + \mathbf{P}^{NL}(\mathbf{r}, t) + \text{complex conjugates}\}$$

that are in turn made up of terms of different physical origins:

$$\begin{aligned} \mathbf{P}^L(\mathbf{r}, t) = {} & \chi^{(1)}(\omega_1) \cdot \mathbf{E}_1(\mathbf{r}) \cdot \exp[i\mathbf{k}_1 \cdot \mathbf{r} - i\omega_1 t] + \chi^{(1)}(\omega_2) \cdot \mathbf{E}_2(\mathbf{r}) \\ & \exp[i\mathbf{k}_2 \cdot \mathbf{r} - i\omega_2 t] + \chi^{(1)}(\omega_3) \cdot \mathbf{E}_3(\mathbf{r}) \exp[i\mathbf{k}_3 \cdot \mathbf{r} - i\omega_3 t] \end{aligned}$$

and

$$\begin{aligned}\mathbf{P}^{NL}(z,t) &= \chi^{(2)}(\omega_3=\omega_1-\omega_2):\mathbf{E}_1\mathbf{E}_2^*\ \exp[i(\mathbf{k}_1-\mathbf{k}_2)\cdot\mathbf{r}-i\omega_3 t]\\ &\quad+\chi^{(2)}(\omega_2=\omega_1-\omega_3):\mathbf{E}_1\mathbf{E}_3^*\ \exp[i(\mathbf{k}_1-\mathbf{k}_3)\cdot\mathbf{r}-i\omega_2 t]\\ &\quad+\chi^{(2)}(\omega_1=\omega_2+\omega_3):\mathbf{E}_2\mathbf{E}_3\ \exp[i(\mathbf{k}_2+\mathbf{k}_3)\cdot\mathbf{r}-i\omega_1 t]\\ &\quad+\text{other nonlinear terms}\end{aligned} \tag{2.18}$$

The key results of the parametric amplification process can be derived by using a one-dimensional scalar model of the wave equation. Furthermore, for the linear amplification regime, the depletion, or the spatial variation, of the amplitude of the "pump-wave" at $\omega_1$, can be neglected; thus, $\mathbf{E}_1(\mathbf{r})$ in this case can be considered to be a fixed parameter $E_p$ equal to the amplitude of the incident pump wave, $E_1(0)$. Effects associated with the depletion of the pump wave due to parametric conversion and back-conversion will be discussed in the following section. Making all these approximations and assuming that the energy and the momentum-conservation conditions, Eqs. (2.1) and (2.2) are satisfied as discussed before, one obtains from Eqs. (2.16)–(2.18) the coupled-wave equations for the complex amplitudes of the signal and idler waves:

$$\left[\frac{\partial^2}{\partial z^2}+\frac{n_2^2\omega_2^2}{c^2}\right]E_2(z)e^{ik_2z}=-\frac{4\pi\omega_2^2}{c^2}d_{\text{eff}}E_pE_3^*e^{ik_2z}, \tag{2.19}$$

and

$$\left[\frac{\partial^2}{\partial z^2}+\frac{n_3^2\omega_3^2}{c^2}\right]E_3(z)e^{ik_3z}=-\frac{4\pi\omega_3^2}{c^2}d_{\text{eff}}E_pE_2^*e^{ik_3z}, \tag{2.20}$$

The corresponding effective Kleinman $d$-coefficient is defined as:

$$d_{\text{eff}}=\sum_{i,j,k=1}^{3}\chi_{ijk}^{(2)}\varepsilon_{2i}\varepsilon_{pj}\varepsilon_{3k} \tag{2.21}$$

where $\varepsilon_{ij}$ is the direction cosine of the $i$-th component of the $\mathbf{E}$-field relative to the $\hat{j}$-axis. $\chi_{ijk}^{(2)}$ refers to the $\chi^{(2)}$ for frequency-mixing process defined in Eq. (2.18). It is equal to 2 times the corresponding Kleinmen's d-coefficient for second harmonic process when dispersion is neglected. In Eqs. (2.19) and (2.20), the indices-of-refraction are related to the linear susceptibilities:

$$\begin{aligned}n_2^2 &= \varepsilon_0+4\pi\chi^{(1)}(\omega_2)\\ n_3^2 &= \varepsilon_0+4\pi\chi^{(1)}(\omega_3)\end{aligned} \tag{2.22}$$

The dispersion in $d_{\text{eff}}$ in Eqs. (2.19) and (2.20) can usually be neglected if all the relevant frequencies are well within the transparency region of the nonlinear material. The effective nonlinear optical coefficient, $d_{\text{eff}}$, takes into account the tensor nature of $\chi^{(2)}$ and the three **E**-fields and must be carefully calculated for each specific case from tabulated values of the relevant components of $\chi^{(2)}$, or the corresponding Kleinman d-tensor. Specific examples of how to calculate the effective $d$-coefficient are given in Chapter 5 below.

Eqs. (2.19) and (2.20) show that the propagation of the signal wave at $\omega_2$ is coupled to the idler wave at $\omega_3$ through the pump wave and the nonlinear susceptibility $d_{\text{eff}}$, and vice versa. The algebra involved in solving the coupled second-order differential equations can be complicated. Fortunately, for optical problems, it is often possible to simplify these equations further by using the so-called "slowly-varying amplitude approximation", in which it is assumed that the change in the complex amplitude of the waves of interest in a distance on the order of one wavelength is much smaller than the complex amplitude itself, or equivalently:

$$\left|\lambda \frac{\partial}{\partial z} E\right| \ll |E| \quad \text{or} \quad \left|\frac{\partial^2}{\partial z^2} E\right| \ll \left|k \frac{\partial}{\partial z} E\right|,$$

which is almost always a good approximation. Making use of this approximation, the second order equations (2.19) and (2.20) can be approximated by the coupled-amplitude equations:

$$\frac{\partial}{\partial z} E_2(z) = i\left(\frac{2\pi k_2}{n_2^2}\right) d_{\text{eff}} E_p E_3^*(z), \tag{2.23}$$

$$\frac{\partial}{\partial z} E_3(z) = i\left(\frac{2\pi k_3}{n_3^2}\right) d_{\text{eff}} E_p E_2^*(z). \tag{2.24}$$

These coupled-amplitude equations have the same form as the coupled-mode equations derived phenomenologically in Section 2.2 on the basis of the repeated difference-frequency picture for the parametric amplification process, but now the proportionality constants are fixed:

$$\xi_2 = i\left(\frac{2\pi k_2}{n_2^2}\right), \tag{2.25}$$

and

$$\xi_3 = i\left(\frac{2\pi k_3}{n_3^2}\right). \tag{2.26}$$

Combining Eqs. (2.23) and (2.24) leads to:

$$\frac{\partial^2}{\partial z^2}E_2(z) = \frac{(2\pi)^2 k_2 k_3 |d_{\text{eff}}|^2 |E_p|^2}{n_2^2 n_3^2} E_2(z), \tag{2.27}$$

which is of the same form as Eq. (2.11) and gives the spatial gain coefficient explicitly:

$$g = \frac{2\pi |d_{\text{eff}}| E_p| \sqrt{k_2 k_3}}{n_2 n_3}. \tag{2.28}$$

Note that $g$ is real and positive. As a numerical example, for a pump intensity of $10^7$ W/cm$^2$ which is easily achievable, a reasonable $d_{\text{eff}}$ value of $5 \times 10^{-9}$ esu, $n_2 \sim n_3 \sim 1.5$, $g$ is approximately 0.4 cm$^{-1}$ at, for example, $\lambda_2\lambda_3 = (700\ \text{nm})^2$. Such a spatial gain value makes a practical optical parametric oscillator quite feasible. The oscillation threshold condition will be given later in Chapter 4.

Solving Eq. (2.27) subject to the boundary condition that there is an input in channel 2 equal to $E_s(0)$ but no input in channel 3 leads to the spatial variations of the complex amplitude of the signal and idler waves:

$$E_2(z) = E_s(0)\ \cosh\ gz, \tag{2.29}$$

and

$$E_3(z) = i\frac{n_2}{n_3}\sqrt{\frac{k_3}{k_2}}E_s^*(0)\ \sinh\ gz. \tag{2.30}$$

Eqs. (2.29) involving the complex amplitudes of the input and output waves shows explicitly the amplitude- and phase-sensitive nature of the amplification of the signal wave.

Since the two channels 2 and 3 are symmetric, one can immediately generalize Eqs. (2.29) and (2.30) to include the contributions of the inputs in both channels:

$$E_s(z) = E_s(0)\ \cosh\ gz + i\frac{n_i}{n_s}\sqrt{\frac{k_s}{k_i}}E_i^*(0)\ \sinh\ gz, \tag{2.31}$$

$$E_i(z) = i\frac{n_s}{n_i}\sqrt{\frac{k_i}{k_s}}E_s^*(0)\ \sinh\ gz + E_i(0)\ \cosh\ gz, \tag{2.32}$$

after changing the notation slightly. These results can be further recast in a particularly useful and completely symmetric form by converting the wave amplitudes to photon flux densities, $\Pi$ [defined as $nE^2c/8\pi h\nu$]:

$$\Pi_s(z) = \Pi_s(0)\ \cosh^2\ gz + \Pi_i(0)\ \sinh^2\ gz, \tag{2.33}$$

and

$$\Pi_i(z) = \Pi_s(0)\ \sinh^2\ gz + \Pi_i(0)\ \cosh^2\ gz, \tag{2.34}$$

These results given in the photon-flux form can be generalized in a rather simple way to include the spontaneous parametric scattering process,[1,4] which is also known as "parametric luminescence"[5] or "parametric fluorescence"[6] in the literature. Spontaneous parametric scattering is analogous to such well-known scattering processes as spontaneous Raman[7] or polariton scattering,[8] or spontaneous Brillouin scattering.[9] The difference is that in the spontaneous parametric scattering process, two photons are generated. In the Raman or Brillouin process, one photon and one phonon are generated. In the polariton process, one photon and one quantum of mixed photon and phonon excitation are generated.

Eqs. (2.33) and (2.34) show that without signal and idler inputs there is no possibility of converting the pump wave into either the signal or the idler wave. This is a consequence of the fact that these equations are derived from classical electromagnetic theory which does not allow for the possibility of spontaneous scattering. Thus, within the frame work of classical electromagnetic theory, the pump wave will go through the nonlinear medium with no probability of spontaneously breaking down into signal or idler photons unless there are signal and idler photons present already. Experimentally, we know that this is not the case, as will be discussed in detail in Chapter 3. Under favorable conditions, there will be copious amounts of spontaneous parametric emission at both the signal and idler wavelengths much more than can be accounted for by mixing of the pump wave with thermal radiation.[10]

Spontaneous parametric emission, as in all spontaneous emission processes, follows from quantization of the electromagnetic fields. Qualitatively, however, spontaneous emission in the signal mode can also be viewed in a semiclassical way as due to the mixing of the pump wave with the zero-point fluctuation in the idler mode.[6,11] The zero-point fluctuation is equal to $h\nu$ per radiation mode per volume for each polarization of the transverse electromagnetic wave. This is analogues to, for example, the spontaneous Raman scattering process. In Raman scattering in solids, for example, the incident photons are converted into Stokes photons and optical phonons. The probability for spontaneous Raman scattering into the Stokes wave can be obtained by assuming one extra phonon per mode per volume,[7] or equivalently the amount of spontaneous Raman emission at the Stokes frequency is equal to that scattered by one optical phonon per mode per volume through stimulated Raman scattering. The analogy here is that the pump wave in the parametric process is analogous to the incident wave in the Raman process, the signal photons are analogous to the Stokes photons, and the idler photons are analogous to the optical phonons. Thus, if spontaneous parametric scattering is to be included, Eqs. (2.33) and (2.34) should have the following modified forms:

$$\Pi_s(z) = \Pi_s(0)\cosh^2 gz + [\Pi_i(0) + \Pi_{zp}]\sinh^2 gz, \tag{2.35}$$

$$\Pi_i(z) = [\Pi_s(0) + \Pi_{zp}]\sinh^2 gz + \Pi_i(0)\cosh^2 gz. \tag{2.36}$$

where $\Pi_{zp}$ represents the equivalent noise input due to the zero-point fluctuations in the signal and idler channels that lead to the spontaneous parametric emission in the idler and signal channels, respectively. In the limit of small $z$, or $\Delta z$, and in the absence of input signal and idler photons, Eq. (2.35) gives the number of signal photons per mode per unit volume, $N_s$, generated per unit time by the spontaneous parametric process:

$$\frac{\partial N_s}{\partial t} \cong \Pi_{zp} g^2 \Delta z, \tag{2.37}$$

where by definition $\Delta N_s \equiv \frac{\Pi_s(\Delta z) n_s}{c}$. It so happens that by assuming one photon per mode per volume or $\Pi_{zp} = c$, Eq. (2.35) and Eq. (2.36) will give the right amount of spontaneous parametric emission in agreement with the results derived from a rigorous theory based upon quantized fields. As will be shown in Section 3.3 below, Eqs. (2.35)–(2.37) form the basis for a semiclassical theory of parametric luminescence that is often quoted in the literature. In the optical domain, the application of these single-mode results are complicated by the fact that open media and, hence, continuous mode spectra are usually involved. For such cases, the single-mode results must be carefully integrated over all the relevant modes in the phase space. Detailed development of such a theory can be found in Refs. 1, 6, and 11.

Spontaneous parametric emission sets the quantum-noise limit of amplitude- and phase-sensitive parametric amplifiers. The equivalent black-body temperature of $h\nu$ per mode per volume in the optical region corresponds to thousands of degrees. Unlike in the microwave region, the parametric amplifier is, therefore, not very useful as a low-noise amplitude- and phase-sensitive amplifier. On the other hand, amplified spontaneous emission can in itself be a useful source of tunable radiation. This is particularly true in the picosecond and femtosecond time domain. In this case, the pump duration is too short for quasi-cw pumping and the repetition rate of available pump sources is often too low for synchronous pumping of parametric oscillators. On the other hand, pumping by short pulses of high peak power can lead to intense single-pass amplified spontaneous emission. As will be discussed in Chapter 6, parametric amplifiers form the basis of a new class of broadly tunable picosecond sources.

## 2.4 LARGE-SIGNAL THEORY OF PARAMETRIC INTERACTION

The discussion above is based on the so-called "small-signal" theory in the sense that the signal and idler intensities are always small compared to the pump intensity so that pump-depletion due to parametric conversion can be neglected. The resulting coupled wave equations or the coupled amplitude equations are linear, with the pump field assumed a constant parameter equal to that of the incident pump wave. In a "large-signal" theory, pump depletion is taken into account explicitly. The basic theory is developed in the classic paper of Armstrong, Bloembergen, Ducuing, and Pershan.[12]

Pump depletion leads to gain saturation in the parametric amplification process, and gain saturation limits the amplitude of the oscillation in the oscillator. In the extreme case, when the pump is completely depleted in a parametric amplifier, the inverse process in which the signal and idler waves recombine into the "pump wave" will take place when the phase-matching condition is satisfied. As long as the phase-matching condition is maintained, energy will continue to flow back and forth between the pump wave and the signal and idler waves.

To include the effects of pump depletion and back-conversion in the theory, the spatial variation of the pump wave must be taken into account in the coupled wave or coupled amplitude equations, (2.23) and (2.24). In addition, a wave equation for the pump field $E_1(z)$ must be introduced. The resulting equations are nonlinear:

$$\left[\frac{\partial^2}{\partial z^2}+\frac{n_p^2\omega_1^2}{c^2}\right]E_1(z)e^{ik_1z} = -\frac{4\pi\omega_1^2}{c^2}d_{\text{eff}}E_2(z)E_3(z)e^{i(k_2+k_3)z}, \tag{2.38}$$

$$\left[\frac{\partial^2}{\partial z^2}+\frac{n_2^2\omega_2^2}{c^2}\right]E_2(z)e^{ik_2z} = -\frac{4\pi\omega_2^2}{c^2}d_{\text{eff}}E_1(z)E_3^*(z)e^{i(k_1-k_3)z}, \tag{2.39}$$

$$\left[\frac{\partial^2}{\partial z^2}+\frac{n_p^2\omega_3^2}{c^2}\right]E_3(z)e^{ik_3z} = -\frac{4\pi\omega_3^2}{c^2}d_{\text{eff}}E_1(z)E_2^*(z)e^{i(k_1-k_2)z}, \tag{2.40}$$

with $E_1(z)$ now also an unknown, as are $E_2(z)$ and $E_3(z)$. It is, however, not practical to solve these coupled nonlinear differential equations directly. Making use of the slowly-varying amplitude approximations, one obtains the coupled amplitude equations:

$$\frac{\partial}{\partial z}E_1(z) = i\left(\frac{2\pi k_1}{n_1^2}\right)d_{\text{eff}}E_2(z)E_3(z)e^{i(k_2+k_3-k_1)z} \tag{2.41}$$

$$\frac{\partial}{\partial z}E_2(z) = i\left(\frac{2\pi k_2}{n_2^2}\right)d_{\rm eff}E_1(z)E_3^*(z)e^{i(k_1-k_2-k_3)z} \tag{2.42}$$

$$\frac{\partial}{\partial z}E_3(z) = i\left(\frac{2\pi k_3}{n_3^2}\right)d_{\rm eff}E_1(z)E_2^*(z)e^{i(k_1-k_2-k_3)z} \tag{2.43}$$

The general solution of these coupled nonlinear complex equations are given in Refs. 12 and 13 in terms of Jacobian elliptic functions. The details of how the solution were arrived at are somewhat involved and will not be repeated here. The pertinent results will be summarized below.

The solution for the normalized signal photon flux density (defined as $u_2 = I_2/\omega_2(I_1 + I_2 + I_3)$] is:

$$u_2^2(\zeta) = u_2^2(0) + u_1^2(0)\left\{1 - sn^2\left[\left(u_1^2(0) + u_2^2(0)\right)^{1/2}|\zeta - \zeta_0|,\right.\right.$$
$$\left.\left.\left(\frac{u_1^2(0)}{u_1^2(0) + u_2^2(0)}\right)^{1/2}\right]\right\} \tag{2.44}$$

for the case of perfect phase-matching and assuming that the idler intensity is zero at the input end [$I_3(0) = 0$] and that the pump intensity is much greater than the signal intensity [$I_1(0) \gg I_2(0)$]. The spatial variation of the pump and idler intensities can be obtained from the so-called Manley-Rowe relations:[14]

$$u_1^2(\zeta) = u_1^2(0) + u_2^2(0) - u_2^2(\zeta), \tag{2.45}$$

$$u_3^2(\zeta) = u_2^2(\zeta) - u_2^2(0), \tag{2.46}$$

which merely reflect the fact that each pump photon that gets converted goes into a signal photon and an idler photon. $\zeta$ in these equations is some sort of a normalized distance from the input end defined as:

$$\zeta = (2\pi/c)d_{\rm eff}[n_1|E_1(0)|^2 + n_2|E_2(0)|^2 + n_3|E_3(0)|^2]^{1/2}[\omega_1\omega_2\omega_3/n_1n_2n_3]^{1/2}z. \tag{2.47}$$

$\zeta_0$ is a parameter that is implicitly defined through the condition that Eq. (2.44) must hold also at $\zeta = 0$, or:

$$1 = sn^2 \left[ \left( u_1^2(0) + u_2^2(0) \right)^{1/2} |\zeta_0|, \left( \frac{u_1^2(0)}{u_1^2(0) + u_2^2(0)} \right)^{1/2} \right] \tag{2.48}$$

With $\zeta_0$ determined from Eq. (2.48), from the known input conditions on $u_1(0)$ and $u_2(0)$, and the boundary condition $E_3(0) = 0$, the spatial dependence of the pump, signal, and idler intensities can now be determined from Eqs. (2.44)–(2.46) and the tabulated values of Jacobian elliptic functions. The pump depletion due to parametric conversion is now fully taken into account.

These results form the basis of the large-signal theory of parametric amplifiers. The Jacobian elliptic function in Eq. (2.44) is an oscillatory function of $\zeta$, which is proportional to the distance from the input end, and describes the conversion of the pump wave into the signal and idler waves and the inverse back-conversion process after complete pump depletion. Additional boundary conditions corresponding to those for the oscillator leads to the theory of optical parametric oscillators including the saturation effect due to pump depletion. It will be discussed in more detail in Section 4.2.

## 2.5 BANDWIDTH OF THE OPTICAL PARAMETRIC PROCESS

The bandwidth of the optical parametric process is determined through the energy-conservation and phase-matching conditions primarily by the linewidth of the pump laser, the angular divergence of the pump wave, and the crystal length.

Consider, for example, the case of Type I collinear phase-matching in a negative uniaxial crystal. As can be seen from Figure 2.2, a finite beam divergence $\Delta\theta_p$ will in general lead to a range of values, $\Delta k_p$, of the magnitude of the propagation vector of the pump beam which will, in turn, lead to a range of values for the sum of the propagation vectors of the signal and idler waves $\Delta(k_s + k_i)$. When the propagation direction of the interacting waves is not along a principal axis of the nonlinear crystal, the linewidth of the signal wave has a first order dependence on the beam divergence. For example, for Type I phase-matching in a negative uniaxial crystal, the bandwidth is:

$$\Delta\omega_s \cong \frac{\omega_p \left( \dfrac{\partial n_p^{(e)}(\theta)}{\partial \theta_p} \right) \Delta\theta_p}{\left| n_s - n_i + \left( \dfrac{\partial n_s^{(o)}}{\partial \omega} \right) \omega_s + \left( \dfrac{\partial n_i^{(o)}}{\partial \omega} \right) \omega_i \right|}, \tag{2.49}$$

from Eqs. (2.1) and (2.2).

In a principal-axis direction, there is no first order change in the index-of-refraction with the phase-matching angle $\theta_p$, or $(\partial n_p^{(e)}(\theta)/\partial\theta_p)$ is zero. The residual linewidth due to the second order terms are usually much smaller than those due to other causes. Phase-matching in such a direction is called "non-critical" phase-matching (NCPM) in the sense that the wavelength of the parametric emission is not sensitive to small changes in the pump beam direction. Under noncritical phase-matching conditions, other mechanisms such as that due to the finite crystal length, or interaction length, may give the dominant contribution to the linewidth. In general the phase-matching condition needs to be satisfied precisely only if the interaction length is infinitely long. If it is finite, this condition needs to be satisfied only to the extent that the waves are nominally in phase within the interaction length. A linewidth can be defined in terms of the change in frequency, $\Delta\omega_{1/2}$, from the precise phase-matching point to where the phase mismatch over the interaction length L is approximately equal to $\pi$; or:

$$\left\{\frac{\partial}{\partial\omega_s}\Delta(k_p - k_s - k_i)\right\}\Delta\omega_s \approx \frac{\pi}{L}, \tag{2.50}$$

and the corresponding full linewidth is then:

$$\Delta\omega_s \cong 2\pi c/L\left|n_s - n_i + \left(\frac{\partial n_s^{(o)}}{\partial\omega}\right)\omega_s + \left(\frac{\partial n_i^{(o)}}{\partial\omega}\right)\omega_i\right|. \tag{2.51}$$

Finally, the photon energy conservation condition, (2.1), clearly shows that any width in the pump frequency, $\Delta\omega_p$, will lead to a corresponding linewidth contribution in the signal frequency $\Delta\omega_s = \Delta\omega_p$.

In conclusion, there are a number of mechanisms that can contribute to the total bandwidth of the parametric process. The principal ones are the pump-beam divergence, finite crystal length, and the pump bandwidth.

## REFERENCES


1. T.G. Giallorenzi and C.L. Tang, *Phys. Rev.*, **166**, 225 (1968).
2. N. Bloembergen and A.J. Sievers, *App. Phys. Lett.*, **17**, 483 (1970).
3. C.L. Tang and P.P. Bey, *J. Quant. Elect.*, **QE-9**, 9 (1973).
4. D. Magde and H. Mahr, *Phys. Rev. Lett.*, **18**, 905 (1967).
5. D.N. Klyshko, *Sov. Phys. – JETP*, **6**, 23 (1967).
6. R.L. Byer and S.E. Harris, *Phys. Rev.*, **168**, 1064 (1968).
7. M. Goeppert-Mayer, *Ann. Phys.*, **9**, 273 (1931).
8. H. Mahr, "*Two-photon absorption spectroscopy*", in *Treatise in Quantum Electronics*, Vol. 1, Nonlinear Optics, Pts. A and B, H. Rabin and C.L. Tang, Eds., New York: Academic Press, 1975.

9. I.L. Fabelinskii, "*Stimulated Mandelstam-Brillouin Process*", in *Treatise in Quantum Electronics*, Vol. 1, Nonlinear Optics, Pts. A and B, H. Rabin and C.L. Tang, Eds., New York: Academic Press, 1975.
10. R.C. Miller and W.A. Nordland, *App. Phys. Lett.*, **10**, 53 (1967).
11. D.A. Kleinman, *Phys. Rev.*, **174**, 1027 (1968).
12. J.A. Armstrong, N. Bloembergen, J. Ducuing, and P.S. Pershan, *Phys. Rev.*, **127**, 1918 (1962).
13. P.P. Bey and C.L. Tang, *J. Quant. Elect.*, **QE-8**, 361 (1972).
14. See, for example, W.H. Louisell, *Coupled Mode and Parametric Electronics*, New York: Wiley, 1960.

# 3. QUANTUM THEORY OF OPTICAL PARAMETRIC PROCESSES

To understand the operations of the optical parametric amplifiers and oscillators which are based upon the stimulated parametric process, a full quantum theory is not necessary. A quantum theory is, however, needed for a rigorous analysis[1,2] of the spontaneous parametric scattering process and a clear understanding of the spontaneous parametric process is important for a number of reasons.

First, because spontaneous parametric emission is proportional to the nonlinear susceptibility squared, it is an alternative method for measuring the nonlinear coefficients of optical materials. In the usual second harmonic generation (SHG) method, because the measured second-harmonic intensity is proportional to the square of the intensity of the fundamental wave, the result is highly sensitive to the temporal and spatial profiles of the incident beam. An advantage of the spontaneous parametric scattering (SPS) method is that the parametric luminescence intensity is linearly proportional to the intensity of the incident wave;[3,4] thus, the corresponding measured value of the nonlinear susceptibility is based upon a relatively accurate power measurement which is not sensitive to the spatial and temporal profiles of the fundamental beam. The theory of the SPS is, however, more complicated. A full quantum field theory is needed to relate the spontaneous emission power to the nonlinear optical susceptibility and the incident pump power.

Second, optical parametric amplifiers form the basis of an important class of tunable picosecond and femtosecond sources. The initial signal leading to the amplified output of such a device comes from spontaneous parametric emission. In addition, the output power of pulsed optical parametric oscillators in the transient regime depends upon the amplified spontaneous emission.

Finally, spontaneous parametric emission is the basic noise source of parametric amplifiers and oscillators. There has recently been a great deal of interest in the possibility of "squeezing" the quantum fluctuations in either the amplitude or the phase of the signal field to below those corresponding to the minimum uncertainty state. By concentrating the signal to be amplified in the squeezed channel, it is possible that the optical parametric amplifier can function as a low-noise amplifier. It is, however, far from clear whether such an amplifier can be made practical, although there are interesting fundamental issues related to squeezing in optical parametric processes that have attracted a great deal of attention.

## 3.1 THE FIELD HAMILTONIAN FOR THE OPTICAL PARAMETRIC PROCESS

To quantize the field, we begin with the field Hamiltonian. For a nonlinear medium, it can be split into two parts:

$$H = H_0 + H_1. \tag{3.1}$$

$H_0$ is the field Hamiltonian corresponding to the medium in the absence of the nonlinearity:

$$H_0 = \frac{1}{8\pi}\int [\mathbf{D}\cdot\mathbf{E} + \mathbf{B}\cdot\mathbf{H}]d\mathbf{r}\,. \tag{3.2}$$

For optical parametric processes, one is usually interested in the transparent region of the optical media. The response of the medium to the fields can then be assumed instantaneous and local. For moderately strong fields, the induced polarization in the medium and, hence, the total Hamiltonian can be expanded in a Taylor series. The corresponding interaction Hamiltonian $H_1$ describing the nonlinear response of the medium is, therefore, of the form:

$$H_1 = \frac{1}{3}\sum_{ijk}\int \chi^{(2)}_{ijk} E_i(\mathbf{r},t)E_j(\mathbf{r},t)E_k(\mathbf{r},t)d\mathbf{r}. \tag{3.3}$$

For spontaneous parametric emission in open media, the fields may not be in isolated modes and the corresponding spectra are generally continuous:

$$\begin{aligned}\mathbf{E}(\mathbf{r},t) &= \frac{i}{2\pi}\int |n_o(\mathbf{k})|^{-1}(h\nu_o)^{1/2}[\hat{a}_o(\mathbf{k})e^{i\mathbf{k}\cdot\mathbf{r}-i\omega_o t} - \hat{a}_o^{+}(\mathbf{k})e^{i\mathbf{k}\cdot\mathbf{r}+i\omega_o t}]\hat{o}(\mathbf{k})d\mathbf{k} \\ &\quad + \frac{i}{2\pi}\int |n_e(\mathbf{k})|^{-1}(h\nu_e)^{1/2}[\hat{a}_e(\mathbf{k})e^{i\mathbf{k}\cdot\mathbf{r}-i\omega_e t} - \hat{a}_e^{+}(\mathbf{k})e^{i\mathbf{k}\cdot\mathbf{r}+i\omega_e t}]\hat{e}(\mathbf{k})d\mathbf{k},\end{aligned} \tag{3.4}$$

where $\hat{a}_e^{+}(\mathbf{k})$ and $\hat{a}_e(\mathbf{k})$ are the creation and annihilation operators, respectively, for extraordinary photons with polarization corresponding to the unit polarization vector $\hat{e}(\mathbf{k})$; the subscript $o$ and the unit polarization vector $\hat{o}(\mathbf{k})$ refer to ordinary waves. Following the usual rules of field quantization, the operators satisfy the following commutation relationships:

$$[\hat{a}(\mathbf{k}), \hat{a}(\mathbf{k}')] = [\hat{a}^+(\mathbf{k}), \hat{a}^+(\mathbf{k}')] = 0 \quad \text{for any } \hat{a} \tag{3.5a}$$

$$[\hat{a}_o(\mathbf{k}'), \hat{a}_o^+(\mathbf{k})] = \delta(\mathbf{k} - \mathbf{k}') \tag{3.5b}$$

$$[\hat{a}_e(\mathbf{k}), \hat{a}_e^+(\mathbf{k}')] = \delta(\mathbf{k} - \mathbf{k}') \tag{3.5c}$$

$$[\hat{a}_e(\mathbf{k}), \hat{a}_o^+(\mathbf{k}')] = 0. \tag{3.5d}$$

The corresponding dispersion relations are:

$$\omega_e n_e(\mathbf{k}) - |\mathbf{k}|c = 0, \tag{3.6a}$$

$$\omega_o n_o(\mathbf{k}) - |\mathbf{k}|c = 0. \tag{3.6b}$$

In terms of this representation of the field, Eq. (3.4), the interaction-free part of the Hamiltonian, $H_0$, becomes:

$$H_0 = \int \left\{ h\nu_o \left[ \hat{a}_o^+(\mathbf{k})\hat{a}_o(\mathbf{k}) + \frac{1}{2} \right] + h\nu_e \left[ \hat{a}_e^+(\mathbf{k})\hat{a}_e(\mathbf{k}) + \frac{1}{2} \right] \right\} d\mathbf{k}. \tag{3.7}$$

For the interaction Hamiltonian, only terms corresponding to the parametric process are of interest. That is, only terms corresponding to the annihilation of a pump photon (assumed extraordinary photon) and the creation of a signal and an idler photon (assumed ordinary photons) and the inverse process need be kept in the interaction Hamiltonian, $H_1$. Upon substituting (3.4) in (3.3), one obtains the modified interaction Hamiltonian relevant to the parametric process:

$$\begin{aligned} H_1' \; = \; -i \iiint \sum_{ijk} \frac{\chi_{ijk}^{(2)} e_i(\mathbf{k}_1) o_j(\mathbf{k}_2) o_k(\mathbf{k}_3)}{n_e(\mathbf{k}_1) n_o(\mathbf{k}_2) n_o(\mathbf{k}_3)} (h^3 \nu_{e1} \nu_{o2} \nu_{o3})^{1/2} \\ \left\{ \hat{a}_e(\mathbf{k}_1)\hat{a}_o^+(\mathbf{k}_2)\hat{a}_o^+(\mathbf{k}_3)\delta(\mathbf{k}_2 + \mathbf{k}_3 - \mathbf{k}_1) \right. \\ \left. -\hat{a}_e^+(\mathbf{k}_1)\hat{a}_o(\mathbf{k}_2)\hat{a}_o(\mathbf{k}_3)\delta(\mathbf{k}_1 - \mathbf{k}_2 - \mathbf{k}_3) \right\} d\mathbf{k}_1 d\mathbf{k}_2 d\mathbf{k}_3 \end{aligned} \tag{3.8}$$

assuming the energy conservation condition, $\omega_{e1} = \omega_{o2} + \omega_{o3}$, or Eq. (2.1), is satisfied; other terms in $H_1$ are of no interest here. The presence of the delta function in the integral ensures that the momentum conservation or phase matching condition, Eq. (2.2), is satisfied.

## 3.2 TRANSITION PROBABILITY AND THE INITIAL AND FINAL STATES

Given the interaction Hamiltonian, the corresponding transition probability from an initial state $|\Phi_{\text{in}}\rangle$ to a final state $|\Phi_{\text{final}}\rangle$ is:

$$W = \lim_{T\to\infty} \frac{1}{VT} \left| \frac{2\pi}{i\hbar} \int_0^T \langle \Phi_{\text{in}} | H_1' | \Phi_{\text{final}} \rangle dt \right|^2 . \tag{3.9}$$

according to the Fermi Golden Rule.

For spontaneous emission, there is no signal and idler photon in the initial state, but there are $N_p$ photons per unit volume on the average in the pump wave mode. The form of the initial state will depend upon whether the pump wave is in a coherent or incoherent state. For an incoherent pump beam, the initial state is a fixed photon number state:

$$|\Phi_{\text{in}}\rangle = |\Phi_{\text{incoh}}\rangle = (2\pi)^{3/2}(N_p!)^{-1/2}[\hat{a}_e^+(\mathbf{k}_p)]^{N_p}|\Phi_0\rangle, \tag{3.10}$$

where $|\Phi_0\rangle$ refers to the vacuum state corresponding to no photon in the medium. The interaction Hamiltonian $H_1'$ leads to the creation of signal and idler photons at the expense of the pump photons:

$$|\Phi_{\text{final}}\rangle = (2\pi)^{3/2}[(N_p - 1)!]^{-1/2}\hat{a}_o^+(\mathbf{k}_s)\hat{a}_o^+(\mathbf{k}_i)[\hat{a}_e^+(\mathbf{k}_p)]^{N_p-1}|\Phi_0\rangle. \tag{3.11}$$

If the pump beam is a coherent wave with an average photon number $\bar{N}_p$ per volume in the pump mode, the initial state is a coherent super-position of fixed photon number states with a Poisson distribution:

$$|\Phi_{\text{in}}\rangle = |\Phi_{\text{coh}}\rangle = (2\pi)^{3/2} \sum_{N=0,1,2,\ldots}^{\infty} \frac{\bar{N}_p^N e^{-\bar{N}_p/2}}{N!^{1/2}} [\hat{a}_e^+(\mathbf{k}_p)]^N |\phi_0\rangle. \tag{3.12}$$

The corresponding final state has one less pump photon but has one signal and one idler photon.

For the stimulated emission process, because it is a *repeated* difference-frequency process, the transition probability Eq. (3.9) based upon the Fermi Golden Rule does not apply. A full quantum theory for the stimulated emission process has yet to be worked out. However, the stimulated parametric process is usually of interest in the context of parametric oscillator or amplifier work where the signal and idler waves are generally coherent and strong enough so that a classical theory based upon Maxwell's equations are adequate, unless questions of noise and fluctuations are involved. The usual practice in dealing with such questions is to separate the problem into two parts. First, the noise source is considered on the basis of a quantum theory as outlined in this chapter or using the heuristic semiclassical theory discussed in Section 2.3. This noise source is then considered as the source term for the amplification process, which can be treated classically as discussed in Section 2.3 also.

## 3.3 SPONTANEOUS PARAMETRIC EMISSION

To obtain the spontaneous emission power per differential solid angle $\Delta\Omega_S$, the transition probability $W$, Eq. (3.9), must be integrated over the distribution function $G(\mathbf{k}_p)$ of the pump-wave in the $\mathbf{k}_p$-space and the appropriate regions in the $\mathbf{k}_s$ and $\mathbf{k}_i$ spaces; the delta functions in $W$ will automatically ensure that only the idler waves that satisfy the phase-matching condition contribute. The total emitted power $P_s$ at the signal wavelength per unit solid angle in the volume $V$ is:

$$\frac{\partial P_s}{\partial \Omega_s} = \iiint WVG(\mathbf{k}_p)h\nu_p\omega_s^2 c^{-3} dk_p d\mathbf{k}_i d\omega_s. \tag{3.13}$$

It can be shown[1] through explicit calculations that the transition probability is independent of the coherence property of the pump beam; that is, both the initial states shown in Eq. (3.10) and Eq. (3.12) give the same final result. In practice, however, incoherent sources are generally not bright enough to start with to serve as an adequate source for parametric scattering to be observable. Using either initial state (3.10) or (3.12), one obtains from Eqs. (3.9) or (3.13) the total integrated spontaneously emitted power per solid angle for an interaction length of $L$:

$$\frac{\partial P_s}{\partial \Omega_s} = \frac{4d_{\text{eff}}^2 h\omega_s^4 \omega_i L P_p}{c^4 n_p n_s n_i \left| n_s - n_i + \dfrac{\partial n_s}{\partial \omega}\omega_s + \dfrac{\partial n_i}{\partial \omega}\omega_i \right|}, \tag{3.14}$$

after carrying out the necessary integrations, where

$$d_{\text{eff}} = \sum_{ijk} \chi_{ijk}^{(2)} e_i(\mathbf{k}_1) o_j(\mathbf{k}_2) o_k(\mathbf{k}_3). \tag{3.15}$$

The details are quite involved but can be found in the literature.[1,2]

It may be of interest at this point to compare the results of the quantum theory, (3.14), with semiclassical theory of parametric luminescence discussed in Section 2.3 following Eq. (2.37). In order to do so, the single mode result given in Eq. (2.37):

$$\Delta N_s = \Pi_{zp} g^2 \Delta z \Delta t = c \cdot g^2 \Delta z \Delta t$$

must be multiplied by the factor:

$$(h\nu_s)\left(\frac{c}{n_s}\right)\left(\frac{\nu_s^2}{c^3} d\nu \, d\Omega\right)$$

to convert it to the emitted power per volume of the nonlinear material over the frequency interval $d\nu$ and differential solid angle $d\Omega$. Replacing $d\nu$ by the bandwidth $\Delta\nu_s$ given by, for example, Eq. (2.49) and expressing the final result in terms of total emitted power for a crystal length of $dz$ instead of $L$ and pump beam power rather than the pump beam intensity, one gets exactly the same result as that derived above on the basis of quantum theory, Eq. (3.14). This establishes the validity of the semiclassical model outlined in Section 2.3, which provides a convenient prescription for estimating the spontaneous emission power without the elaborate details of the quantum theory.

Quantitatively, the spontaneous parametric emission could be 10 orders of magnitude down from the incident power but is easily observable with many conventional pulsed or cw laser sources. The phase-matching condition leads to both collinear interaction and non-collinear interaction. The wavelength of the spontaneous emission will clearly vary with the direction of emission relative to the pump-beam direction. Typically, the spontaneous parametric emission is in the forward direction within a small cone of a few degrees. The color of the spontaneously emitted radiation changes with the emission angle and the whole radiation pattern changes with the orientation of the crystal. A striking example of spontaneous parametric emission is shown in Figures 3.1. Since spontaneous parametric emission is always present and easily observable, it represents both a new tool and a liability.

First, since the frequency and phase-matching conditions for the spontaneous emission are the same as those for the corresponding stimulated parametric process, the tuning characteristics of optical parametric oscillators and amplifiers can be determined from the spontaneous parametric emission prior to the construction of an actual oscillator or amplifier.

Eq. (3.15) shows that the spontaneous emission is linearly proportional to the incident pump power and $d_{\mathrm{eff}}^2$. This relationship can, therefore, be used as the basis for measuring the nonlinear susceptibility of optical materials. As pointed out earlier, the use of spontaneous parametric emission to determine $\chi_{ijk}^{(2)}$ has the important advantage that the spontaneously emitted power is linearly dependent on the incident power which makes it insensitive to the temporal and spatial profiles of the incident beam,

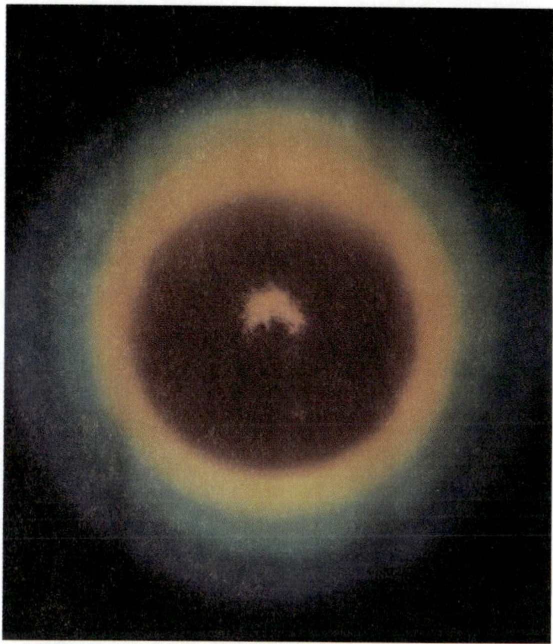

**Fig. 3.1** Spontaneous parametric emission from $\beta$-barium borate pumped at 355 nm.

both of which are more difficult to determine accurately than simply the total power. While there was general agreement on the relative values of the Kleinman $d$-coefficients for most materials, the absolute scale was initially a question of some controversy,[5] because the published values using the SHG and spontaneous parametric scattering methods seem to vary over a range of almost $\sim$ 2. This discrepancy of a factor of $\sim$2 between the two sets of measurements is due to a difference in the definitions of the effective $d$-coefficients for the second-harmonic generation process and the corresponding parametric process (see Eqs. (5.1a) and (5.1b) and the related discussion following these equations in Chapter 5). Taking this factor of 2 into account, the measured results using the SHG and spontaneous parametric scattering methods agree within experimental error. Using a definition of $d_{\text{im}}$ consistent with the usual theory of the SHG process, the numerical result for $d_{31}(2\omega)$ for $LiIO_3$ should be smaller by a factor of 2 than the parametric $d_{31}(\omega_i; \omega_p, \omega_s)$ value given in Refs. 3 and 4, removing any discrepancy between the SHG and spontaneous parametric scattering results. In general, the experimental results based on spontaneous parametric emission should be more accurate for the reasons already discussed, but the theory of second harmonic generation is more straight forward to apply.

Second, since spontaneous emission is always present in the parametric process, it represents an internal source of noise of parametric amplifiers. The spontaneous parametric emission noise associated with the minimum uncertainty state, Eq. (3.12), represents the minimum noise of amplitude- and phase-sensitive parametric amplifiers, because it has its origin in the quantum mechanical uncertainty principle embodied in the commutation relationships, Eqs. (3.5a–d). In the microwave region, the equivalent temperature of the input noise of $h\nu$ per mode per volume is on the order of 0.1°K. In the optical region, the equivalent input noise temperature is in the range of tens of thousands of degrees. Such an optical amplifier is not very useful as a low-noise amplifier. On the other hand, if the signal is in one channel of the amplifier only and the detector after the amplifier is sensitive only to this channel, it is conceivable to use an initial state $|\Phi_{\text{in}}\rangle$ corresponding to a state with the uncertainty or noise concentrated in the channel which the detector is not sensitive to. For the channel the detector is sensitive to, the noise can, thus, be "squeezed" to below that corresponding to the minimum uncertainty state without violating the uncertainty principle. This is the concept of "squeezing". With squeezing, the noise figure of an amplitude-sensitive or a phase-sensitive, but not both, amplifier can, in principle, be improved beyond that corresponding to the minimum noise figure of an amplitude- and phase-sensitive amplifier. How practical such an approach will be is still not known. Unlike in the microwave region, the primary interest in optical parametric amplifiers is, therefore, to use it as a source of powerful tunable radiation. This is particularly true in the picosecond or shorter time domain.

## REFERENCES


1. T.G. Giallorenzi and C.L. Tang, *Phys. Rev.*, **166**, 225 (1968).
2. D.A. Kleinman, *Phys. Rev.*, **174**, 1027 (1968).
3. A.J. Campillo and C.L. Tang, *App. Phys. Lett.*, **16**, 242, 537 (1970). The expression for $\Delta\varepsilon_1$ given in Eq. (3) is in terms of a $d_{\text{im}}$ for parametric process. Expressing in terms of a $d_{\text{im}}$ consistent with that of conventional SHG usage, Eq. (3) in this reference would read $\Delta\varepsilon_1 = 8\pi \sum_{ijk} d_{im}(2\omega) e_i o_j o_k = 8\pi d_{31}(2\omega)\sin(\theta_p + \rho)$. This leads to a measured $d_{31}(2\omega)$ value of $\sim 9\times 10^{-19}$ esu in agreement with the SHG results. See the discussion following Eqs. (5.2a) and (5.2b) in Chapter 5.
4. M.M. Choy and R.L. Byer, *Phys. Rev. B*, **14**, 1693 (1976).
5. D.A. Roberts, *J. Quant. Elect.*, **28**, 2057 (1992).

# 4. OPTICAL PARAMETRIC OSCILLATORS

As a source of broadly tunable coherent radiation, the parametric device can be an oscillator or a generator of amplified fluorescence. The form of the device will depend on the available nonlinear crystals and pump sources and the design trade-offs. One of the basic limitations is that the average pump power is generally on the order of a few watts given the current state of the technology.

This average-power barrier puts a constraint on the combinations of energy/pulse, peak power, pulse width, and pulse repetition rate that can be achieved in various wavelength range. Thus, if pulses of tens of mJ or more and of several nanoseconds long are needed, the pulse repetition rate is generally limited to less than a few hundred Hz. With such low repetition pump rates, the duration of the pump pulse must be significantly longer than the optical parametric oscillator (OPO) cavity round-trip time in order for the OPO signal wave to build up from spontaneous parametric noise level in a "quasi-cw" oscillation. With currently available sources such as the Nd:YAG laser with its second, third, and fourth harmonics of $\sim$10 nanosecond long pulses as the pump and nonlinear crystals such as BBO and KTP, it is possible to construct such quasi-cw pumped OPO's. In fact, this type of OPO is the first successful commercial OPO.

For nonlinear optical applications in the picosecond and femtosecond time domain where high peak power is needed, quasi-cw pumping is not possible because the short pump pulse duration does not allow multiple passes in the OPO cavity. Synchronous pumping, where the short pump pulse travels in synchronism with the signal pulse, is necessary. In this case, if multi-$\mu$J is needed, the pulse repetition rate has to be of the order of 100 KHz or lower. At this repetition rate, it is not possible to sustain oscillation in an OPO cavity of a reasonable length. The corresponding optical parametric device tends to be an optical parametric amplifier seeded by a broad band radiation which is often the spontaneous parametric emission coming from a pumped nonlinear optical crystal.

Finally, high-repetition rate ($\sim$100 MHz or higher) synchronously-pumped continuous pulse-train OPO's can now yield hundreds of mW average power from the visible to the mid infrared down to the 20 femtoseconds range. With intracavity or extracavity doubling and sum-frequency generation, the tuning range can be further extended to the near-uv. Many different types of such femtosecond OPO's have already been demonstrated and more are yet to come. The main advantages of oscillators over amplified parametric luminescence are higher conversion efficiency, narrower linewidth, and without the need for a separate device to generate the parametric luminescence. Thus, if available, an oscillator is often preferred.

We begin our consideration of optical parametric devices with the oscillators. In principle, the optical parametric oscillator can consist simply of a nonlinear optical crystal and two mirrors forming a Fabry-Perot cavity, as shown earlier in Figure 1.3. Tuning is achieved by simple rotation of the crystal. Practical OPOs are not much more complicated than that. There are two basic types of OPO: the singly-resonant oscillator (SRO) in which only the signal wave is resonated while the idler wave is lost from the cavity on each passage, and the doubly-resonant oscillator (DRO) in which both the signal and idler waves are resonated. The SRO has a higher threshold for oscillation, but is inherently stable. The DRO has a lower threshold for oscillation, but tends to be highly unstable with the output often consisting mainly of random spikes much like some of the flashlamp-pumped solid state lasers that are not Q-switched, unless special cavity-stabilization schemes are employed.

## 4.1 OSCILLATION THRESHOLD CONDITIONS

For parametric oscillators in general, the oscillation threshold condition is simply that the total round-trip gain over the interaction length is equal to the loss, just as in ordinary lasers.

### 4.1.1 Singly-Resonant Oscillators (SRO)

In the singly-resonant OPO, the Fabry-Perot cavity mirrors reflect at only one of the down-converted wavelengths, which is by convention somewhat arbitrarily designated as the signal wave in the case of oscillators. The cavity mirrors in the SRO, therefore, typically reflect the signal wave and transmit the idler and pump waves. In some cases where the pump wavelength is in the uv, because of mirror coating damage considerations, the pump beam might not be brought in and out of the cavity through the cavity mirrors but by separate beam-steering mirrors that reflect the pump beam but transmit the signal and idler waves, as shown in Figure 4.1. Apart from these practical considerations, all OPO's have the same basic configuration.

Given the basic oscillator configuration in the form of a nonlinear optical crystal in a Fabry-Perot cavity, the oscillation threshold condition can be obtained by applying the proper boundary conditions. With amplitude reflectivities of the mirrors at the input ($z = 0$) and the output ($z =$ crystal or effective interaction length $L$), $r_1$ and $r_2$, respectively, the boundary conditions are:

$$E_s(0) = r_1 r_2 E_s(L) \quad \text{and} \quad E_i(0) = 0. \tag{4.1}$$

Applying these to the solutions, Eqs. (2.29) and (2.30), of the wave equations, (2.19) and (2.20):

$$E_s(L) = E_s(0) \cosh gL = r_1 r_2 E_s(L) \cosh gL, \quad \text{or} \quad 1 = r_1 r_2 \cosh gL. \tag{4.2}$$

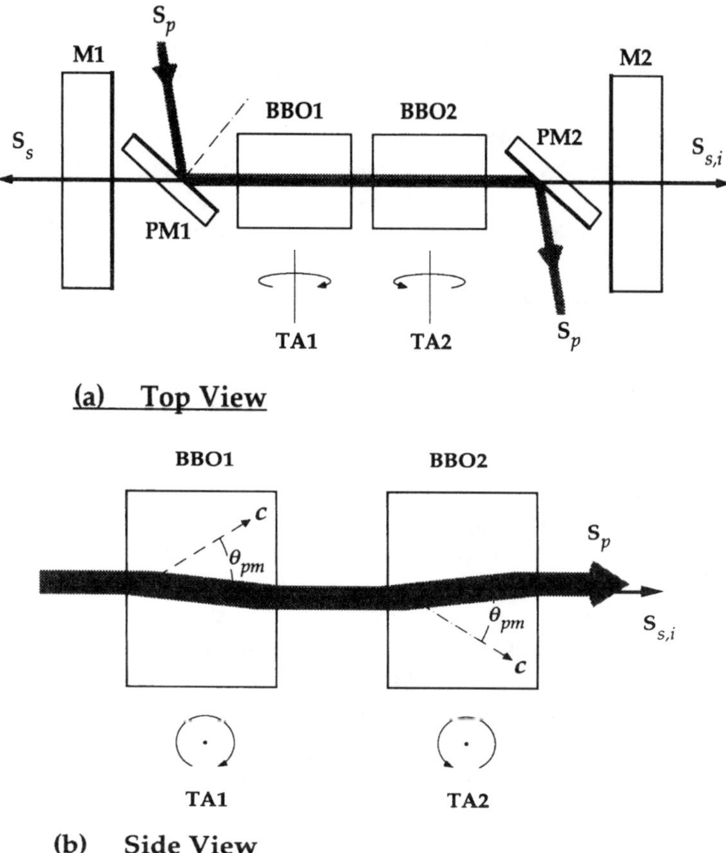


**Fig. 4.1** Schematic of BBO OPO with two-crystal and beam-steering arrangement. **TA** = tuning axis, **PM** = pump mirror, **M** = signal mirror, **S** = Poynting vector and ***c*** = crystal optic axis. (See Section 6.1.1 for details).

For the SRO, there is no feedback of the idler wave. Substituting the expression for gain $g$, Eq. (2.28), in Eq. (4.2) gives the general oscillation threshold condition for the SRO. A useful form of the condition is that in the limit of small $gL$ or the high mirror-reflectivity case. In this limit and assuming the two mirror reflectivities are the same, $r_1 = r_2 \equiv R^{1/2}$, Eq. (4.2) gives the oscillation threshold condition of a singly resonant OPO:

$$\left(I_p L^2\right)^{(\mathrm{SRO})}_{\mathrm{threshold}} \cong \frac{n_p n_s n_i \lambda_s \lambda_i c}{128\pi^5 d^2} \frac{\left(1 - R^2\right)}{R^2}, \tag{4.3}$$

where the $\lambda$'s are the free-space wavelengths. As a numerical example, for a nonlinear optical crystal such as $\beta$-barium borate (BBO) pumped by the 3rd harmonic of a Nd-YAG laser, an $(I_pL^2)_{\text{threshold}}$ value of $3.5 \times 10^6$ $W$ is obtained for an $R$ value of 90%. An interaction length of 1 cm in BBO crystal is easily achievable giving a threshold pump intensity of ~3.5 MW/cm$^2$ or on the order of 30 mJ/cm$^2$ for a typical 8 nsec pulse from a wavelength-tripled pulsed Nd-YAG laser, which is not difficult to achieve. With the large variety of pump sources and nonlinear crystals now available, singly-resonant OPO's are becoming a commonly available source of broadly tunable coherent radiation.

The effective parametric interaction length may, however, be shorter than the physical length of the nonlinear crystal. The limiting length is often the so-called walk-off length if it is shorter than the crystal length. This is because both ordinary and extraordinary waves are always involved in the phase-matching condition. For ordinary waves, the phase velocity or the wave vector is in the same direction as that of the Poynting vector. For the extraordinary wave, this is in general not the case (see Figure 4.2) except along the principal axes. As a result, the beam envelopes of the ordinary and extraordinary waves will walk-off from each other if the phase-velocities are matched in the same direction. This could well limit the effective interaction length. There are situations where the optimum condition requires noncollinear phase-matching in order to maximize the interaction length by aligning the group velocities.[1,2] As will be discussed in connection with the femtosecond OPO's, when curved cavity mirrors are used and the oscillation direction is not restricted by the cavity in one particular direction as in Fabry-Perot cavities consisting of plane mirrors, there are situations where the OPO will automatically optimize itself by oscillating in the direction where the Poynting vectors are aligned corresponding to noncollinear interaction as far as the phase-velocities are concerned.

### 4.1.2 Doubly-Resonant Oscillators (DRO)

In the doubly-resonant oscillator, the mirrors reflect both the signal and the idler waves. The corresponding boundary conditions are:

$$E_s(0) = r_{s1}r_{s2}E_s(L), \quad \text{and} \quad E_i(0) = r_{i1}r_{i2}E_i(L), \tag{4.4}$$

instead of (4.1). Applying these boundary conditions to Eqs. (2.31) and (2.32):

$$E_s(L) = R_sE_s(L)\cosh gL + i\frac{n_i}{n_s}\sqrt{\frac{k_s}{k_i}}R_iE_i^*(L)\sinh gL \tag{4.5}$$

$$E_i^*(L) = -i\frac{n_s}{n_i}\sqrt{\frac{k_i}{k_s}}R_sE_s(L)\sinh gL + R_iE_i^*(L)\cosh gL \tag{4.6}$$

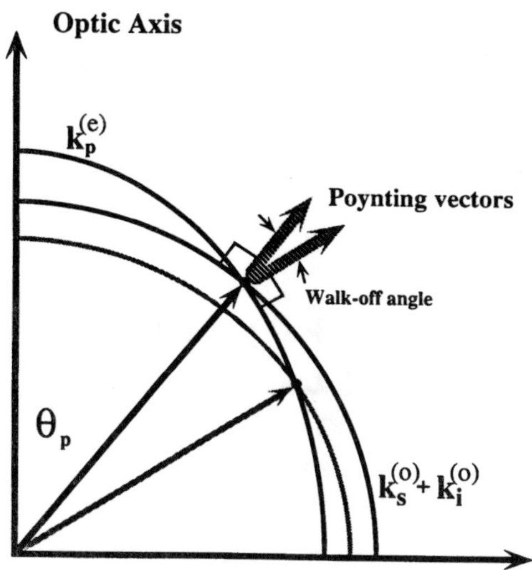


**Fig. 4.2** The direction of the Poynting vector is normal to the $k$-vector surface. Thus, for phase-matched waves, the Poynting vectors of the ordinary and extra-ordinary waves are in different directions, leading to the walk-off effect.

yields the threshold condition in the case of small $gL$ and assuming $R_s = R_i = R$:

$$\frac{\left(I_p L^2\right)^{(\mathrm{DRO})}_{\mathrm{threshold}}}{\left(I_p L^2\right)^{(\mathrm{SRO})}_{\mathrm{threshold}}} \cong \frac{1-R}{1+R}. \tag{4.7}$$

Since $R$ can be well over 90%, for example, (4.7) shows that the threshold pump intensity required for the DRO can be significantly smaller than that for the SRO. There is, however, a very important trade-off consideration. The stability of the DRO is usually not good. In fact, the outputs of most cw or quasi-cw type of DROs are often in the form of random spikes as those typically seen in ordinary flash-lamp pumped ruby or other solid state lasers without Q-switching.

The explanation for the instability in the DRO is that it is more sensitive to cavity length fluctuations because there are two cavity-resonance conditions to be satisfied simultaneously leading to the so-called clustering effect. This is shown schematically in Figure 4.3 where the resonances for the idler modes is shown as a function of $\omega_p - \omega_i = \omega_s$. Due to dispersion in the intracavity optical components, the spacing between these resonances are in general not the same as that of the resonances for the signal wave. The OPO can only oscillate at where the resonances from both sets are within the widths of these resonances. Such coincidence tends to occur in

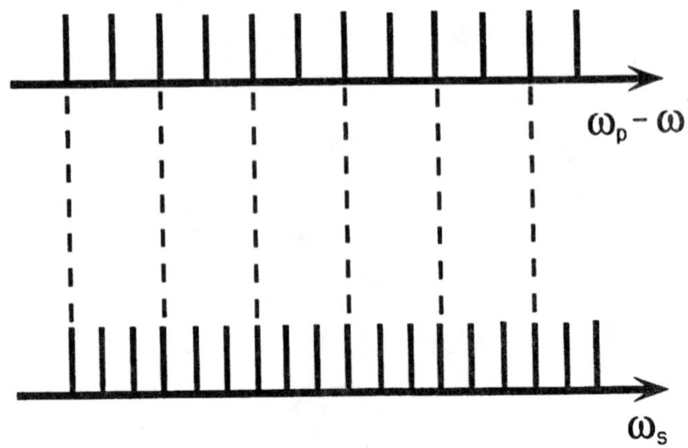


**Fig. 4.3** Signal and idler resonances of a doubly-resonant OPO showing the clustering effect.

well separated clusters as shown in Figure 4.3. Fluctuations in the cavity length, for example, would cause the oscillating clusters to jump among the possible signal and idler modes leading to strong intensity fluctuations. Stabilization of the cavity-length by feedback-control can reduce this type of instability and lead to low oscillation threshold for DRO's.[3] Detailed experimental and numerical modeling of SRO and DRO cw OPOs have recently been reported.[4,5] In particular, Yang *et al.* have examined the issue of OPO frequency stability in the SRO-to-DRO transition region and shed light on the compromise between oscillation threshold and frequency stability as the mirror reflectivity for the idler wave is varied.[4]

To get around this frequency stability problem, researchers at St. Andrews University[6] used a three mirrors, dual-cavity oscillator to resonate the signal and idler waves separately. By servo-locking the two cavities, frequency stable DRO outputs continuously tunable over a ~0.4 GHz range have been demonstrated. The relatively low ~200 mW DRO threshold observed for the device is particularly encouraging, since it suggests that with lower loss optical elements and improved *trichroic* mirror reflectivity control, cw OPO's can become a practical technology in applications ranging from optical frequency standards to spectroscopy.

## 4.2 LARGE SIGNAL THEORY OF PARAMETRIC OSCILLATORS

Because the basic parametric interaction is a single-step three-photon process, at the end of such a process the medium returns to the initial state and there is no dissipative loss to the medium involved. Whatever is lost from the coherent pump beam is converted into useful photons at the signal and idler frequencies. Thus, the OPO can be a highly efficient device once it is above the threshold for oscillation. In contrast, in an optically pumped three-level laser, for example, three single-step quantum-transitions are involved. There is dissipative loss associated with each transition because of the finite lifetime or linewidth with its implied relaxation process associated with each discrete energy level of the laser medium. In addition, one of the transitions is often a nonradiative relaxation process; as a result, a part of the input pump energy is dissipated into heat.

For efficient parametric conversion, it is implied that a substantial fraction of the pump power is depleted and converted into signal and idler photons. The theory presented in the previous section is a small-signal theory which neglects the pump-depletion effect. For estimating the quantum efficiency of optical parametric oscillators, a large-signal theory is, therefore, needed.

The basic large-signal theory for parametric amplifiers was outlined in Section 2.4. The key results are Eqs. (2.44)–(2.48), which give the spatial variations of the pump, signal, and idler intensities. For an SRO, the oscillator boundary conditions are:

$$u_s^2(\zeta = 0) = R_1 R_2 u_s^2(\zeta = L) \tag{4.8}$$

and

$$u_i(\zeta = 0) = 0. \tag{4.9}$$

Let $\zeta_L$ be the value of $\zeta$ when $z = L$ in the definition of $\zeta$, or from Eq. (2.47):

$$u_p(0)\zeta_L = (32\pi^3/c^3)^{1/2} I_p^{1/2}(0) d_{\text{eff}} L; \tag{4.10}$$

it has the meaning, therefore, of a normalized interaction length. Substituting Eq. (4.8) and Eq. (4.10) in Eq. (2.44) gives

$$u_s^2(0) = \left(\frac{R_1 R_2}{1 - R_1 R_2}\right) u_p^2(0) \left\{1 - sn^2\left[\left(u_p^2(0) + u_s^2(0)\right)^{1/2} |\zeta_L - \zeta_0|, \left(\frac{u_p^2(0)}{u_p^2(0) + u_s^2(0)}\right)^{1/2}\right]\right\}. \tag{4.11}$$

The only two unknowns are the parameter $\zeta_0$ and $u_s^2(\zeta = 0)$, or the intracavity signal intensity at the input end $\zeta = 0$. These two unknowns can be found by solving Eq. (4.11) and Eq. (2.48), or:

$$1 = sn^2\left[\left(u_p^2(0) + u_s^2(0)\right)^{1/2} |\zeta_0|, \left(\frac{u_p^2(0)}{u_p^2(0) + u_s^2(0)}\right)^{1/2}\right] \tag{4.12}$$

simultaneously. Because of the complexity of these equations, numerical methods are usually required. Once $u_s^2(\zeta = 0)$ is known, important physical properties of SRO such as the output power above the threshold for oscillation, optimum coupling, residual transmitted pump power, etc. can be found. The oscillation-threshold condition corresponds to the value of $u_p^2(0)$ when $u_s^2(\zeta = 0) \approx 0$. The result is of course the same as that of the small-signal theory given in Chapter 2.

Examples of some key numerical results are shown in Figures 4.4–4.6. These results illustrate a number of the important features of the large-signal behavior of the SRO:

*1) Signal Intensity Inside the Cavity* – Figure 4.4 shows the ratio of normalized intracavity signal photon flux $u_s^2(0)$ to $u_p^2(0)$ as a function of the product of pump intensity and the nonlinear crystal length $L$ squared normalized to the threshold value, or $[I_p(0)L^2]/[I_p(0)L^2]_{\text{th}}$, for various values of mirror transmission $T = 1 - R_1R_2$. A key feature is that the intracavity signal photon flux can be much higher than that of the pump photon flux; the intensity ratio must of course be reduced by the frequency ratio $\omega_s/\omega_p$. Since optical damage in the nonlinear crystal or intracavity coatings is often an important consideration in the design of high power parametric oscillators, this fact has important practical implications. The intracavity damage level could be set by the signal intensity in the cavity rather than that of the pump. Another possible consequence of this fact is that, in femtosecond optical parametric oscillators, strong intracavity signal intensity can lead to chirping in the generated signal pulse due to self-phase modulation in the nonlinear crystal.

*2) Quantum Efficiency* – A quantum efficiency can be defined for the parametric oscillator:

$$\eta = \frac{(1 - R_2)u_s^2(\zeta_L)}{u_p^2(0)} = \left(\frac{1 - R_2}{R_1R_2}\right)\frac{u_s^2(0)}{u_p^2(0)}. \tag{4.13}$$

From the $sn^2$-function dependence in $u_s^2(0)$ given in Eq. (4.11), it can be shown that at $\zeta_L = \zeta_0$, the quantum efficiency is at maximum and

$$\eta_{\max} = \frac{1 - R_2}{1 - R_1R_2} \leq 1.$$

The maximum quantum efficiency for SRO is, therefore, 100% when $R_1 = 1$.

Figure 4.5 gives the ratio of the pump power-$\zeta_0^2$ product at maximum efficiency, or $[u_p(0)\zeta_0]^2$, to that at the threshold, or $[u_p(0)\zeta_L]_{\text{th}}^2$, for various values of $T$. Note that this ratio is on the order of 2 to 2.5 with the limiting value of $(\pi/2)^2$ as first derived by Kreuzer[7] on the basis of the approximation which neglects the intracavity spatial variation of the intensity. This is a very important point in practical OPO design. It shows that, for maximum efficiency, the output coupling should be such that the pump power is at about 2.5 times the threshold for oscillation.

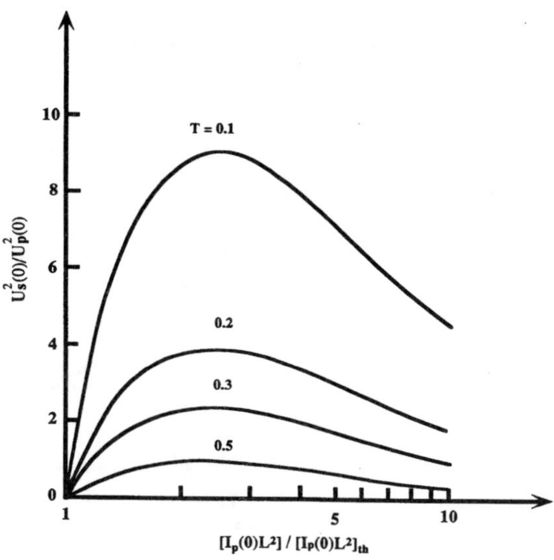


**Fig. 4.4** Ratio of the signal-to-pump flux densities as a function of the normalized pump intensity at various output coupling loss values $T = 1 - R_1R_2$, where $R_1R_2$ are the intensity reflectivities of the cavity mirrors.

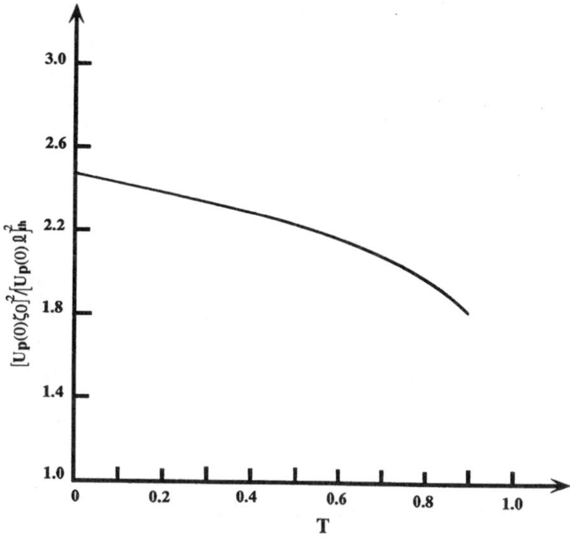


**Fig. 4.5** Ratio of the normalized pump at maximum efficiency to that at the threshold as a function of the transmission of the output coupling mirror.

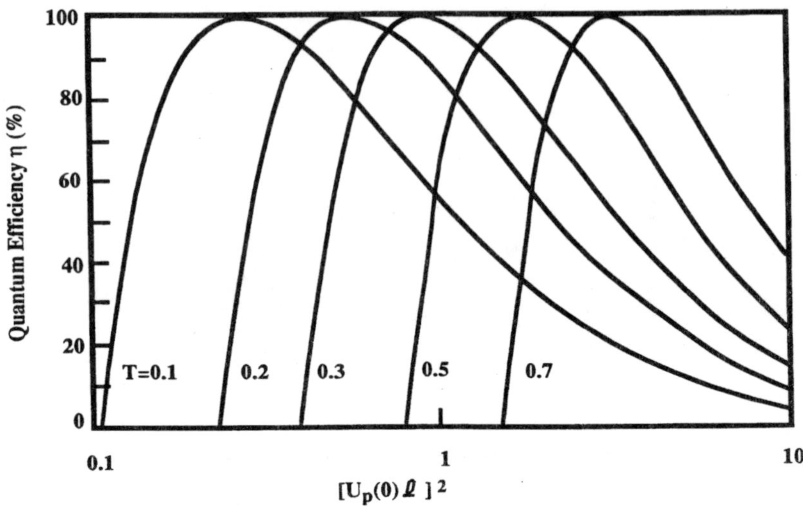


**Fig. 4.6** Quantum efficiency of singly-resonant optical parametric oscillator (SRO) as a function of pump power for various values of cavity loss parameter $T$.

Figure 4.6 shows the quantum efficiency $\eta$ versus the pump power-$\zeta_L^2$ product for various values of $T$ with $R_1 = 1$. It shows again that the maximum efficiency can be 1 and where it occurs. It is interesting to note that at higher pump power levels, the output coupling $T$ should also be higher to achieve maximum quantum efficiency. These results also show the consequence of "back-conversion" [ i.e. the reconversion of signal and idler waves back into the pump wave]. As Figure 4.6 shows that for a given pump power the quantum efficiency actually decreases as the mirror reflectivity increases beyond optimum coupling. This back-conversion effect has recently been seen in femtosecond OPO's.

*3) Threshold Condition* – Although Figure 4.6 also gives the threshold condition for oscillation [when $\eta = 0$], near the threshold the small-signal approximation should be good enough and the results given in Section 4.1 above would be completely adequate.

## 4.3 PRACTICAL EFFICIENCY CONSIDERATIONS

Unlike in the lasers, the optical parametric oscillator based on the three-photon parametric process converts photons at one wavelength to photons at two other wavelengths with basically no dissipative loss. The conversion process can be very efficient sufficiently above the threshold for oscillation. In lasers, nonradiative transitions representing dissipative losses are usually involved in establishing the population-inversion required for achieving gain in the laser medium. There is a limit in how efficient the process can be.

The maximum theoretical quantum efficiency of parametric conversion either to the signal or idler wave in cw optical parametric oscillators with the pump-wave propagating only in the forward direction in the OPO cavity is 100% in the case of singly-resonant oscillator as shown in Section 4.2 above and 50% in the case of doubly-resonant oscillator under idealized condition. The difference in the two cases is that, in the latter case, the signal and idler waves are present in both the forward and backward propagating directions while the idler wave is absent in the backward direction in the former case. Since there is no incident pump wave in the backward direction, in the case of DRO, the co-propagating signal and idler waves can recombine into a backward-propagating wave at the pump frequency satisfying the same phase-matching condition. This generated wave at the pump frequency in the backward direction then becomes a reflected pump-wave. It has been shown by Siegman[8] that the corresponding maximum reflection coefficient is 50%. Thus, in the case of DRO, 50% of the incident pump photons will be reflected at maximum efficiency and all the remaining incident forward-travelling pump-wave will be completely converted into signal and idler waves, giving rise to a 50% maximum total quantum efficiency. In the case of SRO, there is no idler wave in the backward direction in the OPO cavity to recombine with the backward-travelling signal wave. Thus, 100% of the forward-travelling pump photons can in principle be converted to the signal and idler photons under the optimum condition. The corresponding signal power efficiency is of course less than the quantum efficiency, because the signal-to-pump frequency ratio is less than unity.

Note that the phase-matching condition for parametric down-conversion from the pump wave to the signal and idler waves is exactly the same as that for the inverse sum-frequency process. Thus, in an optical parametric amplifier, as the pump-wave is totally depleted and converted to the signal and idler waves, the latter will recombine into a wave at the pump-frequency and the energy will oscillate back and forth between the waves as long as the phase coherence of the waves is maintained in the process. Optimum conversion for optical parametric amplifiers requires terminating the process at the point of 100% quantum conversion.

For practical optical parametric oscillators, a number of factors conspire to make the efficiency lower than the maximum possible efficiency under idealized condition. Apart from the pump power level and interaction length requirements, the more important factors are intracavity losses, the transverse intensity variation, and the finite pulse duration. A plane-wave theory for OPO conversion efficiency was first developed by Kreuzer[7] using the approximation that the signal intensity is constant

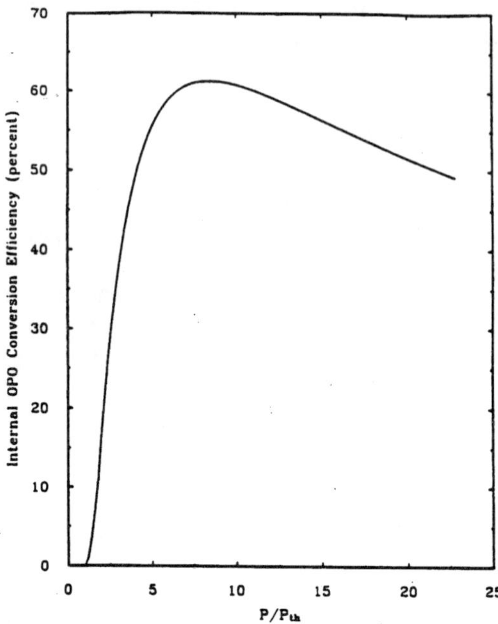


**Fig. 4.7** Calculated efficiency curve of OPO based upon the Byer-Brosnan modification (Ref. 10) of Bjorkholm's Gaussian-profile model (Ref. 11).

spatially inside the OPO cavity, which is only valid when the mirror reflectivities are high and the conversion efficiency or the pump-depletion is small and the signal wave has essentially a constant amplitude throughout the cavity. The theory was further extended by Bjorkholm to include a Gaussian transverse field variation.

If the conversion is not small, which in practical OPO's can be on the order of 30% or more, even if the transverse spatial variation is neglected, there can still be strong longitudinal spatial variations in the pump, signal, and idler intensities due to pump depletion and amplification of the signal and idler waves toward the output coupling mirror. In this case, the problem becomes fully nonlinear and the large signal theory given in Section 4.2 must be used. The spatial variations of the fields are described not by sinusoidal functions but Jacobian elliptic functions. As shown in detail in the previous Section 4.2 and in Ref. 9, the full nonlinear problem based on the plane-wave approximation is still quite complicated. Furthermore, it takes into account the longitudinal spatial variations due to the pump-depletion and the finite mirror reflectivity, but not the transverse spatial variation of the intensities of all the fields. A full large-signal theory taking into account also the transverse field variation is yet to be developed.

For practical pulsed OPO's in the nanosecond range, the actual oscillator efficiency is, however, most often limited by the number of passes that the signal and idler waves can make during the pump pulse duration. An approximate formula proposed by Brosnan and Byer[10] for the transient case is to modify the efficiency result of Bjorkholm[11] by a multiplication factor $\sim (1 - P_{\text{th}}/P)$, where $P$ is the pump power and $P_{\text{th}}$ is the pump power at the threshold for oscillation. Thus, the shorter the pump pulse, the higher is the pump power required to reach oscillation and the lower the efficiency at the same pump power above the threshold. A numerical curve characterizing the OPO SRO total power (signal plus idler) efficiency for the transient case based upon the Brosnan-Byer modification of Bjorkholm's result is shown in Figure 4.7. As can be seen, the efficiency of practical pulsed OPO's can indeed be very high. The practical efficiency of an OPO is often limited by the optical damage, however.

## 4.4 PARAMETRIC OSCILLATOR TUNING AND LINEWIDTH CONSIDERATIONS

As discussed in Chapter 2, the tuning characteristics of parametric amplifiers and oscillators are completely determined by the energy conservation and phase-matching conditions, Eqs. (2.1) and (2.2). They are, therefore, identical to those for the corresponding spontaneous parametric emission.

The basic spectral width of the parametric process is due to the pump bandwidth, the pump beam divergence, or the crystal length. In practice, it can vary from a few Å to hundreds of Å near the degenerate point. As the bandwidth formulas Eqs. (2.49) and (2.51) show, the bandwidth can be especially large near the degenerate point where $\omega_s$ and $\omega_i$ and, hence, $n_s$ and $n_i$ are equal. A large potential bandwidth can be an advantage or disadvantage depending on applications. It is obviously a disadvantage when building a oscillator for high spectral resolution applications. It is a distinct advantage when building short pulse optical parametric amplifiers or oscillators, where a large bandwidth is needed to accommodate very short pulses.

The oscillator linewidth is further reduced from that of the basic parametric amplification process. For practical OPO's under pulsed or quasi-cw condition, the oscillator linewidth can be estimated qualitatively from the basic bandwidth $\Delta\omega_s$ of the parametric gain in the OPO. It is approximately equal to the single-pass linewidth divided by the square-root of the number of passes permitted by the pump duration. For moderately high gain, the single-pass bandwidth is approximately the basic bandwidth of the parametric process, $\Delta\omega_s$ (e.g. Eq. (2.49)).

In practice, without employing any special line-narrowing schemes, the linewidth for typical OPO's is on the order of a few Å or greater. To reduce the linewidth to less than an Å, two schemes are commonly used: grating reflector and injection locking.

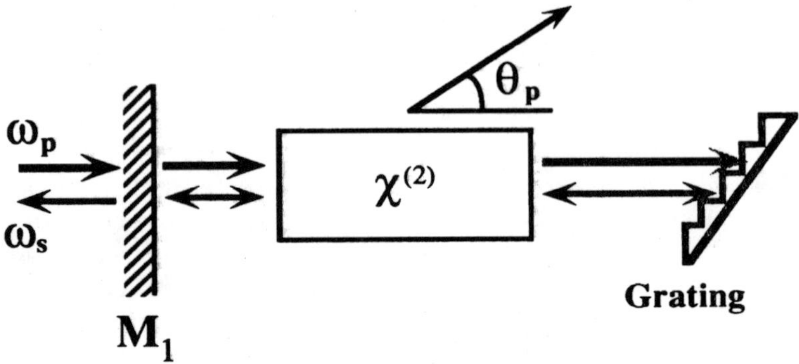


**Fig. 4.8** Schematic of optical parametric oscillator with line-narrowing grating retro-reflector in the Littrow configuration.

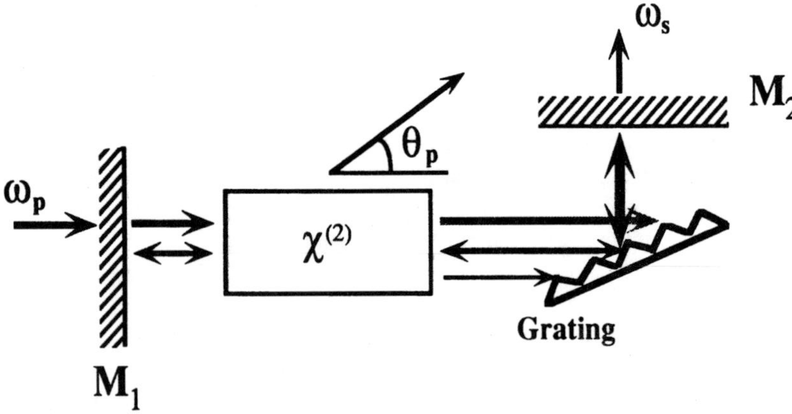


**Fig. 4.9** Schematic of optical parametric oscillator with line narrowing grating reflector in the Littman configuration.

The use of a grating reflector in Littrow-configuration or the so-called Littman-configurations as shown in Figures 4.8 and 4.9, respectively, generally reduces the linewidth by about an order of magnitude. This linewidth reduction is often achieved at the expenses of lower efficiency and higher oscillation threshold due to the possibly lower reflectivity of the grating arrangement, and due to the increased cavity length which reduces the number of passes of the signal pulse in the cavity for a given pump pulse duration. For simultaneous narrow linewidth and high power applications, the most commonly used method is to use a narrow linewidth master OPO to injection-lock a high power OPO to achieve the final high power output. A linewidth reduction a little over an order of magnitude is also possible through injection locking of pulsed OPO's.

The use of Type II phase-matched parametric interaction generally give narrower line width than Type I interaction would. In Type II interaction, the signal and idler waves are of different polarizations which leads to different dispersion of the waves even when the two wavelengths are the same.

Specific examples of the implementation of these schemes will be given in Chapter 6.

## 4.5 OPTICAL DAMAGE

Optical damage in parametric oscillators is a most important issue. Damage can occur both in the nonlinear crystal and in the mirror coatings. In fact, damage in the crystals was one of the major factors that impeded the development of the optical parametric oscillators for many years. With the variety of high power pump sources currently available, reaching oscillation threshold is not a major problem for a number of nonlinear crystals. On the other hand, the achievable efficiency of the OPO's is often still limited by the damage threshold of the crystals and the mirror coatings rather than the available pump laser energy. The damage issue in connection with commonly used crystals is discussed in Chapter 5.

Damage in mirror coatings remains a serious problem especially if the pump wavelength is in the uv. In the simplest pumping configuration for an OPO, the pump beam is introduced into and passed out of the cavity through the end mirrors of the cavity as shown schematically in Figure 1.3. Thus, the end mirrors must transmit at the pump wavelength and reflect at the signal wavelength. For the uv spectral region, because the damage threshold of mirror coatings tend to be lower for uv-transmitting and visible reflecting mirrors than for uv-reflecting but visible- and ir-transmitting mirrors, it is sometimes advisable to avoid coupling the pump light in through the Fabry-Perot cavity end mirrors of the OPO oscillator. Instead, the pump light can be coupled into and out of the cavity through a pair of intracavity "pump-steering" mirrors as shown schematically in Figure 4.1. These mirrors transmit at the signal wavelength and reflect the pump beam at an angle to the signal wave propagation direction.

In considering optical damage in OPO, it should also be remembered that, as shown in Figure 4.4, the intracavity signal intensity can be much higher than the incident pump intensity.

To avoid damages, it is obviously important that the beam profile be smooth and there is no hot spots in the pump beam. Hot spots can also lead to poor collimation of the pump beam which will in turn lead to unintended broadening of the OPO output linewidth. Spatial filtering or injection locking of the pump beam to a high-quality seed beam are commonly used methods to smooth out the beam profile. Because OPO is scalable, to avoid mirror and crystal damages due to high average intensities, one can always make the beam size larger at high power levels.

## REFERENCES

1. D.C. Edelstein, E.S. Wachman, C.L. Tang, *App. Phys. Lett.*, **54**, 1728 (1989).
2. L.A.W. Gloster, Z.X. Jiang, and T.A. King, *J. Quant. Elect.* (to be published).
3. D. Lee and N.C. Wong, *J. of Opt. Soc. Am. B*, **10**, 1659 (1993).
4. S.T. Yang, R.C. Eckardt and R.L. Byer, *J. Opt. Soc. Am. B*, **10**, 1684 (1993).
5. R.C. Echardt, C.D. Nabors, W.J. Kozlovsky and R.L. Byer, *J. Opt. Soc. of Am. B*, **8**, 646 (1991).
6. F.G. Colville, M.J. Padgett and M.H. Dunn, *Appl. Phys. Lett.*, **64**, 1490 (1994).
7. L.B. Kreutzer, in *IERE Proc. Joint Conf. Lasers and Optoelectronics*, London, England, 1969.
8. A.E. Siegman, *App. Optics*, **1**, 739 (1962).
9. P.P. Bey and C.L. Tang, *J. Quant. Elect.*, **QE-8**, 361 (1972).
10. S.J. Brosnan and R.L. Byer, *IEEE J. of Quant. Elect.*, **QE-15**, 415 (1979).
11. J.E. Bjorkholm, *J. Quant. Elect.*, **QE-7**, 109 (1971).

# 5. MATERIALS, PROPERTIES AND CHARACTERIZATION

While the importance of OPO as a useful source of tunable coherent radiation was clear from the beginning and a variety of theoretical and experimental studies were carried out, the realization of practical OPO technology was hampered by the lack of suitable nonlinear optical crystals. The development of nonlinear optical materials is unfortunately a slow and tedious process, with the discovery of truly useful new optical materials often a matter of chance. This lack of progress in practical development coupled with over-optimism on the part of some early enthusiasts eventually led to considerable skepticism about the prospects of the optical parametric oscillator as a useful device. In the meantime, progress in the search of new nonlinear optical crystals was slowly being made during the past decade, leading to many new nonlinear optical crystals (e.g. $\beta$-$BaB_2O_4$, $LiB_3O_5$, L-arginine phosphate, MgO:$LiNbO_3$, $KTiOPO_4$ and its isomorphs). This, along with the steady improvement in the crystal quality of efficient nonlinear crystals such as $KNbO_3$, $AgGaS_2$, $AgGaSe_2$, $ZnGeP_2$ and $Tl_3AsSe_3$, has led to the recent rapid progress in optical parametric oscillator technology. To better appreciate the importance of these new crystals, we need to understand the material property requirements for nonlinear crystals suitable for use in optical parametric oscillators.

## 5.1 MATERIAL REQUIREMENTS

Simply put, the desirable properties of OPO crystals are: broad transparency, high nonlinearity, sufficient birefringence for phase matching, high damage threshold, good temperature stability, low optical loss, chemical inertness and ease of sample fabrication (which includes the availability of large homogeneous crystals and compatibility with crystal polishing and coating technology). Of these properties, the optical nonlinearity is perhaps the easiest to obtain, as evidenced by the large number of nonlinear organic crystals that has been reported in the past decade.[1] This is due in part to Kurtz and Perry's powder second harmonic generation test,[2] which has allowed a large number of new compounds be screened quickly and semiquantitatively for their nonlinearity. Until recently, measurement of phase matching properties, such as refractive indices, SHG cutoff, angular acceptance and temperature bandwidth, of a new material required the tedious and often expensive process of single crystal growth. The microcrystal characterization technique recently introduced by Velsko[3] promises to speed up the search of new materials with better overall characteristics significantly. In this technique, a small crystal sphere of $\leq 1\ mm^3$ is obtained by air-tumbling. The small sphere is then mounted on the tip of a glass fiber and submerged in appropriate index-matching fluid for linear and nonlinear optical properties assessment. With this approach, fairly accurate measurement of these properties can be obtained before valuable resources are committed to growing bulk crystals of a new material. However, there is as yet no quick way to evaluate the optical damage resistance and the optical losses of new nonlinear materials without bulk crystals of sufficient size, and we must resort to the time-consuming process of growing high quality single crystals suitable for direct measurement.

As the primary interest in OPO is the convenient generation of broadly tunable coherent radiation, the nonlinear crystals must of course be optically transparent over a sufficiently broad output wavelength range. Since all parametric processes are based on a short wavelength pump laser, the transparency range of the nonlinear crystal should overlap the output wavelengths of available pump lasers such as the Nd:YAG and its harmonics (1.064, 0.532, 0.355 & 0.266 $\mu$m), Ti-sapphire (0.7–1.0 $\mu$m), alexandrite (700–800 nm), holmium and erbium lasers ($\sim$2–3 $\mu$m) and the excimer lasers (e.g. 308 nm of XeCl). This requirement may seem obvious. Yet, it is useful to point out that during the past decade, a great deal of research effort in the development of infrared materials such as $AgGaSe_2$ and $ZnGeP_2$ has been devoted to improving the optical transparency of these materials at the pump laser wavelengths.[4,5] Similarly, the requirement that crystal transparency must overlap the pump, signal and idler wavelengths, and the implicit requirement of broad output wavelength tunability rule out an entire class of aromatic organic crystals for most practical OPO applications.

Crystal transparency also determines directly its phase-matching properties, and thus the tunability of the OPO. For material with broad transparency, the dispersion is small and less birefringence is needed to achieve phase-matching. A smaller birefringence is particularly desirable in OPOs as it reduces the problem of beam-'walkoff' which often limits the effective interaction length. For parametric generation of wavelengths near the crystal absorption cutoff (uv or infrared), large birefringence will be needed.

Low material dispersion is also desirable in the parametric generation of ultrashort pulses. Just as crystal birefringence can lead to spatial beam-walkoff, group velocity dispersion can cause the temporal pulse separation between the pump and the resonant signal as they propagate through the nonlinear crystal. This limits the effective interaction length among the three waves and often contributes to temporal pulse broadening.[6] For example in a Ti-sapphire pumped femtosecond OPO using the crystal $KTiOPO_4$ (KTP), the group velocity mismatch between the $\sim$780 nm pump and the *ordinary* 1.56 $\mu$m signal causes the pump to lag behind the signal by $\sim$150 fsec for each mm of travel in the crystal. Figure 5.1 illustrates this group velocity mismatch in KTP under the collinear phase-matching geometry. Coincidentally, the *extraordinary* wave's group velocity in this case is very close to that of the pump, suggesting that crystals longer than a few mm may be used if spatial beam-walkoff

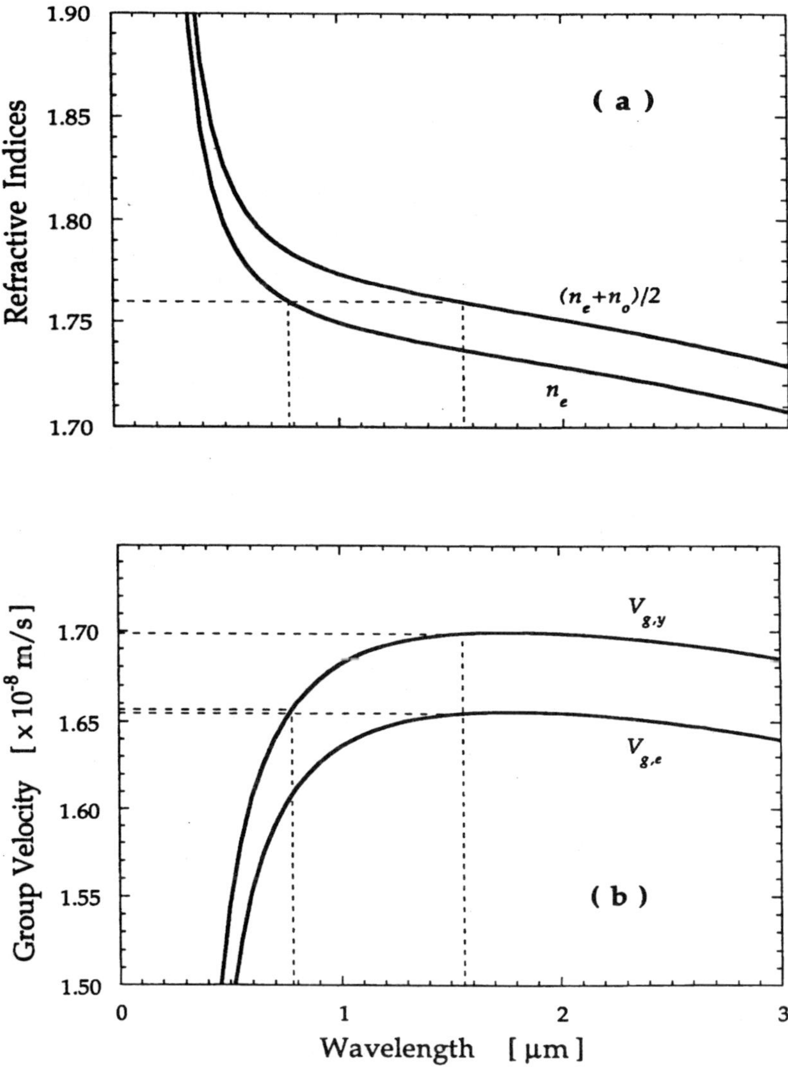


**Fig. 5.1** (a) Type II collinear phase velocity matching in the *xz* plane of $KTiOPO_4$. The 0.78 $\mu$m Ti:$Al_2O_3$ pump laser enters the crystal as extraordinary wave, and the signal and idler outputs are degenerate at 1.56 $\mu$m. With subpicosecond pulses, the group velocity ($V_g$) mismatch among the pump, signal and idler for this process is illustrated in (b).

can be compensated for. The 'optimal' crystal length will be a compromise between the need for high parametric gain and minimal pulse width, which of course depends on the OPO's desired output characteristics. As in the case of nanosecond-pulsed OPOs, design consideration often leads to trade-offs among the desirable properties of the nonlinear medium. The dynamics of synchronously pumped picosecond and femtosecond OPOs are complex,[7,8] and there is as yet no clear understanding of the material requirements for the nonlinear crystals in these applications.

It can be argued that conventional material selection criteria may not be applicable for OPO/OPA operating in the femtosecond regime. To avoid pulse broadening, the nonlinear medium is often limited to no more than 1–2 mm long. This short interaction length can render bulk crystal limitations unimportant. For instance, a 1 mm long $KNbO_3$ crystal has a respectable temperature bandwidth of $\sim$3°C. Along with its large nonlinearity and high tuning rate – hence broad spectral bandwidth – $KNbO_3$ may prove to be particularly useful in ultrafast applications. Similarly, weak optical absorption loss at the output wavelengths may be less detrimental to the operation of OPO's using these thin crystals. For instance, in the synchronously pumped femtosecond KTP OPO reported in reference 6, the peak intensity of the resonating *signal* infrared pulses of a KTP OPO was estimated at $\sim$1–10 GW/cm$^2$. With such high intensity, non-phase matched second harmonic generation can become the dominant loss mechansim, and the efficient generation of 650 nm femtosecond pulses via the large non-phasematchable $d_{33}$ coefficient of KTP ($\sim$18 pm/V) have already been reported.[6,9] While this non-phasematched harmonic output from the OPO was viewed as advantageous by its inventors, for low intracavity loss application, crystals with smaller non-phasematchable $d_{ij}$ coefficients, such as $LiB_3O_5$, are preferred.[10] Lastly, little is known about the optical damage mechanism in the femtosecond time scale, and new damage threshold measurements are clearly needed. It is therefore premature to compare existing nonlinear materials for this application.

In the nanosecond regime, optical damage of nonlinear crystals, such as $LiNbO_3$, urea and $AgGaS_2$, has impeded early development of the optical parametric oscillator. The maximum optical damage threshold of transparent materials is of course its laser-induced dielectric breakdown limit.[11] Owing to crystal imperfection and other intrinsic material properties, this dielectric breakdown threshold is often not realized in real crystals. For instance, it is well known that congruent melting $LiNbO_3$ has limited usefulness in high power frequency conversion application because of its low photorefractive damage threshold. Similarly, the photochromic reduction of $Ti^{4+}$,[12] more commonly referred to as gray-tracking, in KTP remains an important material problem which reduces its reliability in very high power application near its 350 nm band edge. Extrinsic material properties, such as surface damage caused by flaws introduced during polishing, are also key material problems which vary greatly among different nonlinear crystals. For reliable OPO operation, Byer has suggested limiting the pump field to wavelengths approximately one-half the material bandgap.[13] The availability of a complementary set of high quality nonlinear crystals, such as BBO, KTP and $AgGaS_2$, with different absorption cutoffs should make this a realistic design approach to provide continuous coverage of the uv to mid-infrared spectrum.[14]

The optical damage requirement for materials used in parametric generation is more stringent than for second harmonic generation. The pump power is at the shortest wavelength and is often near the uv absorption edge of the crystal. Near the uv absorption cutoff, optical damage threshold of typical dielectrics are found experimentally to decrease with laser wavelength. In parametric down conversion, the combination of higher intensity and short pump wavelength poses stringent optical damage resistant requirement on the nonlinear optical crystal and cavity optics. In addition, depending on the oscillator design, the resonant signal wave inside the oscillator cavity can build up to fluence levels exceeding those of the pump, as already discussed in Section 4.2,[9] and crystal damage caused by the resonantly enhanced signal wave has been reported.[13] Indeed, among the improved characteristics offered by the new nonlinear materials, their significantly higher optical damage thresholds (e.g $\sim$2 GW/cm$^2$ for KTP *vs*. 140 MW/cm$^2$ for $LiNbO_3$;[13,15] and $\sim$10 GW/cm$^2$ for BBO *vs*. 180 MW/cm$^2$ for urea[16,17]) are crucial to bringing about the recent advances in OPO technology.

Though less important in single-pass optical parametric generation (OPG), low optical loss in the nonlinear crystal are essential in reducing oscillation threshold of OPO, especially in low power, continuous wave devices.[18] Optical loss can be due to optical scattering or absorption. For most nanosecond or longer OPO applications, long ($>$1 cm) crystals are often needed to compensate for the limited nonlinearity of existing crystals. Thus, even weak absorption, such as vibration overtones...etc., can lead to significantly higher oscillation thresholds. In a singly resonant OPO, this can be overcome by resonating the less lossy of either the blue or the red branch (relative to the wavelength degeneracy) of the OPO tuning curves. Nevertheless if efficient output coupling is important, even weak overtone absorption must be avoided. For example, in KTP the weak multiphonon absorption[19] centered at $\sim$3.5 $\mu$m makes the generation of 3–4 $\mu$m radiation in this crystal inefficient, and has so far limited its usefulness in this spectroscopically important region.[20]

For material with small temperature phase-matching bandwidth, optical absorption can cause crystal heating and lead to reduced conversion efficiency through thermally-induced dephasing among the interacting pump, signal and idler waves. In extreme cases involving high average power, severe laser induced heating can cause thermal loading of the crystal, resulting in catastrophic crystal failure. Although these problems can be managed[21] for crystals with high thermal conductivity and/or small thermal expansion coefficient, for design simplicity, materials with low optical absorption and broad temperature bandwidths are preferred.

Low optical scattering in crystals represents another challenging material problem. The exact cause of optical scattering varies, ranging from microscopic inclusions of bubbles,[22] precipitates or microcrystallites,[4] to refractive index inhomogeneity caused by strain,[23] composition fluctuation[24] or impurity (e.g. flux) incorporation.[25] Besides increasing the optical loss, microscopic scattering centers often serve as initiation sites for optical damage. Since the presence of only a few inclusion centers within the interaction volume is sufficient to lower the damage resistance of a crystal regardless of its size, the growth of inclusion free material becomes significantly more difficult as one scales to longer interaction length and larger crystal aperture. In addition, scattering can lead to drastic reduction in conversion efficiency by distorting the wave fronts. The infrared crystals $Tl_3AsSe_3$ and $AgGaS_2$ are just two of the numerous examples in which the reduction of scattering centers correlates directly with higher conversion efficiency.[26,27] Similarly from the above discussion, it is clear that the practical use of 'damage-resistant' MgO:$LiNbO_3$ crystals in OPO applications will depend greatly on the ability to grow these doped crystals free of undesirable compositional striation.[28–30]

To achieve efficient energy conversion from the pump laser into the signal and idler waves, the crystal nonlinearity must be sufficiently large to provide efficient coupling among these waves. While it is in theory possible to compensate for a small nonlinearity by increasing the crystal length, currently factors such as spatial beam-walkoff and crystal growth limitation make the use of crystal longer than 3 cm impractical. The nonlinear coupling constant, $d_{\mathrm{eff}}$ in Eq. (2.23), depends on the crystal's nonlinear susceptibility tensor $\mathrm{d}_{ij}$ and the crystal's birefringence which governs its phase matching properties. For a given phase matching configuration – i.e. Type I or II, the effective coupling constants are obtained from the projection of the electric fields **E** onto the susceptibility tensor (c.f. Eq. 2.18).[31] The effective d-coefficient for parametric processes consistent with the corresponding coupled wave equations for the signal and idler waves of the form shown in Eqs. (2.19) and (2.20) is:

$$d_{\mathrm{eff}}^{(\mathrm{Parametric})} = \sum_{i,j,k=1}^{3} 2d_{ijk}(\omega_i;\,\omega_p\omega_s)\varepsilon_i(\omega_i)\varepsilon_j(\omega_p)\varepsilon_k(\omega_s) \tag{5.1a}$$

where $\varepsilon_i(\omega)$ is the direction cosine of the field component at $\omega$ relative to the $\hat{\imath}$-axis [see also Eq. (2.21)]. Note that, to keep the form of the wave equation for the second-harmonic generation (SHG) process the same as Eqs. (2.19) or (2.20), the corresponding effective $d$-coefficient is:

$$d_{\mathrm{eff}}^{(\mathrm{SHG})} = \sum_{i,j,k=1}^{3} d_{ijk}(2\omega;\,\omega,\omega)\varepsilon_i(2\omega)\varepsilon_j(\omega)\varepsilon_k(\omega) \tag{5.1b}$$

Thus, taking into account the Bloembergen symmetry condition $d_{ijk}(2\omega;\,\omega,\omega) = d_{jik}(\omega;\,2\omega,\omega)$, the effective $d$-coefficients for the parametric process and the corresponding SHG process given by Eqs. (5.1a) and (5.1b), respectively, differ by a factor of 2. This difference of a factor of 2 in the expressions for the effective $d$-coefficients for SHG and the corresponding parametric process, which holds even in the degenerate case ($\omega_i = \omega_s$), is sometimes overlooked in the literature and it accounts for much of the confusion and discrepancies in the numerical values of the $d$-coefficients determined by the SHG and the spontaneous parametric scattering techniques. In some papers, the effective $d$'s for the two processes are defined the same. In which case, the polarization source terms in the corresponding wave equations must differ by a factor 2.

The evaluation of $d_{\mathrm{eff}}$ is best illustrated with an actual example: the $d_{\mathrm{eff}}$ for collinear, Type I down conversion process in $\beta$-$BaB_2O_4$ (BBO). Since BBO is negative uniaxial, Type I phase matching is achieved by propagating the pump as the only extraordinary wave ($e \to o + o$). Figure 5.2 depicts the polarizations of the pump, signal and idler waves relative to the piezoelectric axes. We have used the standard orientation[32] for the point group *3m*, i.e. X$\perp$ $m$ and Z$\|\mathbf{P}_s$ the spontaneous polarization. In BBO, the piezoelectric {X,Y,Z} axes coincide respectively with the optical dielectric $\{x, y, z\}$ axes. The component of the nonlinear polarization oscillating at the idler frequency is given explicitly by, in the esu system:

$$\begin{pmatrix} \mathbf{P}_x(\omega_i) \\ \mathbf{P}_y(\omega_i) \\ \mathbf{P}_z(\omega_i) \end{pmatrix} = 2 \begin{pmatrix} 0 & 0 & 0 & 0 & \mathbf{d}_{15} & -\mathbf{d}_{22} \\ -\mathbf{d}_{22} & \mathbf{d}_{22} & 0 & \mathbf{d}_{15} & 0 & \\ \mathbf{d}_{31} & \mathbf{d}_{31} & \mathbf{d}_{33} & 0 & 0 & 0 \end{pmatrix} \begin{pmatrix} \mathbf{E}_x(\omega_p)\mathbf{E}_x(\omega_s) \\ \mathbf{E}_y(\omega_p)\mathbf{E}_y(\omega_s) \\ \mathbf{E}_z(\omega_p)\mathbf{E}_z(\omega_s) \\ \mathbf{E}_y(\omega_p)\mathbf{E}_z(\omega_s) + \mathbf{E}_z(\omega_p)\mathbf{E}_y(\omega_s) \\ \mathbf{E}_x(\omega_p)\mathbf{E}_z(\omega_s) + \mathbf{E}_z(\omega_p)\mathbf{E}_x(\omega_s) \\ \mathbf{E}_x(\omega_p)\mathbf{E}_y(\omega_s) + \mathbf{E}_y(\omega_p)\mathbf{E}_x(\omega_s) \end{pmatrix} \tag{5.2}$$

where we have used the contracted notation $d_{lm}$, with $l = \{x, y, z\} = \{1, 2, 3\}$ and $m = \{xx, yy, zz, yz, xz, xy\} = \{1, 2, 3, 4, 5, 6\}$, for the third rank tensor $d_{ijk}$.[33] For Type-I phase-matching in a negative uniaxial crystal where the pump wave is an extraordinary wave and the signal and idler waves are ordinary waves, for example, the corresponding effective nonlinear coupling is given by:

$$\frac{E_o(\omega_i)}{E_e(\omega_p)E_o(\omega_s)} = d_{\mathrm{eff}}^{(\mathrm{Parametric})} = \sum_{i,j,k=1}^{3} = 2d_{ijk}(\omega_i; \omega_p\omega_s)a_i(\omega_i)b_j(\omega_p)a_k(\omega_s) \tag{5.3}$$

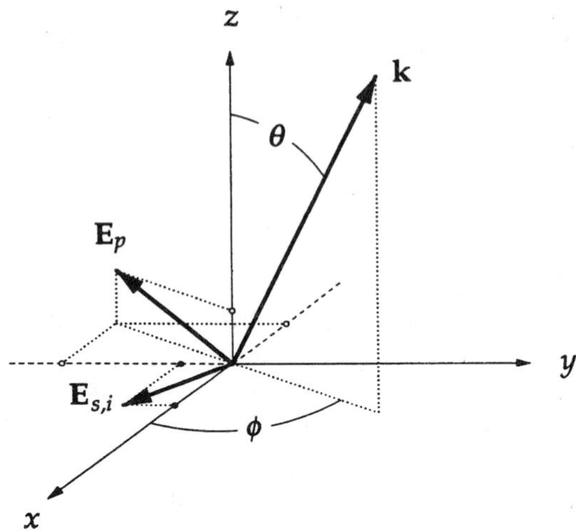


**Fig. 5.2** Polarization of the pump ($\mathbf{E}_p$), signal and idler ($\mathbf{E}_{s,i}$) for Type I phase matching ($e \rightarrow o + o$) in $\beta$-$BaB_2O_4$. (•) and (∘) denote the projection of $a_j$ and $b_j$ respectively. By convention, the piezoelectric X-axis coincides with the optical $x$-axis, and is chosen to be perpendicular to the crystallographic mirror plane $m$ (see Figure 5.12).

where $a_{i,k}(\omega)$'s and $b_j(\omega)$'s are the direction cosines of the ordinary and extraordinary fields, **E** and **P**:

$$\begin{pmatrix} a_x \\ a_y \\ a_z \end{pmatrix} = \begin{pmatrix} \sin\phi \\ -\cos\phi \\ 0 \end{pmatrix}; \text{ and } \begin{pmatrix} b_x \\ b_y \\ b_z \end{pmatrix} = \begin{pmatrix} -\cos\theta\cos\phi \\ -\cos\theta\sin\phi \\ \sin\theta \end{pmatrix}. \tag{5.3a}$$

Evaluating the summation in Eq. (5.3) leads to the effective $d$-coefficient for Type I parametric generation in BBO:

$$d_{\text{eff}}^{(\text{Parametric})}(\theta, \phi) = 2[d_{31}\sin\theta - d_{22}\cos\theta\sin 3\phi]. \tag{5.4}$$

The more general case of non-collinear phase matching can be obtained in a similar fashion. In BBO, $d_{31}$ is approximately 20 times smaller than $d_{22}$ and does not contribute to the effective nonlinearity.[34] Thus, to maximize the nonlinearity, the crystal is cut for light propagation in the $yz$-plane (with $\phi = \pm 90°$). In the case of MgO:$LiNbO_3$, which is also negative uniaxial with $3m$ symmetry, the two nonlinear susceptibilities are comparable ($d_{31}/d_{22} \sim 2.1$). The most efficient crystal cut is now obtained from $\{\theta_{\text{pm}} > 0°, \phi = 90°\}$.[35] Table 5.1 summarizes the expressions for the effective nonlinearity for SHG in uniaxial crystals.[33]

**Table 5.1 (a)** Effective nonlinear coupling constant ($d_{eff}^{SHG}$) for positive uniaxial crystals.[33] The cubic point groups 23 and $\bar{4}$3m are not listed as they have zero birefringence.

| Crystal Point Group | Positive Birefringence $n_e > n_o$ Type I; $o \to e+e$ | Type II; $o \to o+e$ |
|---|---|---|
| 6; 4 | $-d_{14} \sin 2\theta$ | $d_{15} \sin\theta$ |
| 622; 422 | $-d_{14} \sin 2\theta$ | 0 |
| 6mm; 4mm | 0 | $d_{15} \sin\theta$ |
| $\bar{6}$m2 | $d_{22} \cos^2\theta \cos 3\phi$ | $-d_{22} \cos\theta \sin 3\phi$ |
| 3m | $d_{22} \cos^2\theta \cos 3\phi$ | $d_{15} \sin\theta - d_{22} \cos\theta \sin 3\phi$ |
| $\bar{6}$ | $(d_{11} \sin 3\phi + d_{22} \cos 3\phi) \cos^2\theta$ | $(d_{11} \cos 3\phi - d_{22} \sin 3\phi) \cos\theta$ |
| 3 | $(d_{11} \sin 3\phi + d_{22} \cos 3\phi) \cos^2\theta - d_{14} \sin 2\theta$ | $(d_{11} \cos 3\phi - d_{22} \sin 3\phi) \cos\theta + d_{15} \sin\theta$ |
| 32 | $d_{11} \cos^2\theta \sin 3\phi - d_{14} \sin 2\theta$ | $d_{11} \cos 3\phi \cos\theta$ |
| $\bar{4}$ | $(d_{14} \cos 2\phi - d_{15} \sin 2\phi) \sin 2\theta$ | $-(d_{15} \cos 2\phi + d_{14} \sin 2\phi) \sin\theta$ |
| $\bar{4}$2m | $d_{14} \sin 2\theta \cos 2\phi$ | $-d_{14} \sin 2\phi \sin\theta$ |

**Table 5.1 (b)** Effective nonlinear coupling constant ($d_{eff}^{SHG}$) for negative uniaxial crystals.[33]

| Crystal Point Group | Negative Birefringence $n_e < n_o$ Type I; $e \to o+o$ | Type II; $e \to e+o$ |
|---|---|---|
| 6; 4 | $d_{31} \sin\theta$ | $d_{14} \sin\theta \cos\theta$ |
| 622; 422 | 0 | $d_{14} \sin\theta \cos\theta$ |
| 6mm; 4mm | $d_{31} \sin\theta$ | 0 |
| $\bar{6}$m2 | $-d_{22} \cos\theta \sin 3\phi$ | $d_{22} \cos^2\theta \cos 3\phi$ |
| 3m | $d_{31} \sin\theta - d_{22} \cos\theta \sin 3\phi$ | $d_{22} \cos^2\theta \cos 3\phi$ |
| $\bar{6}$ | $(d_{11} \cos 3\phi - d_{22} \sin 3\phi) \cos\theta$ | $(d_{11} \sin 3\phi + d_{22} \cos 3\phi) \cos^2\theta$ |
| 3 | $(d_{11} \cos 3\phi - d_{22} \sin 3\phi) \cos\theta + d_{31} \sin\theta$ | $(d_{11} \sin 3\phi + d_{22} \cos 3\phi) \cos^2\theta + d_{14} \sin\theta \cos\theta$ |
| 32 | $d_{11} \cos 3\phi \cos\theta$ | $d_{11} \sin 3\phi \cos^2\theta + d_{14} \sin\theta \cos\theta$ |
| $\bar{4}$ | $-(d_{31} \cos 2\phi + d_{36} \sin 2\phi) \sin\theta$ | $[(d_{14} + d_{36}) \cos 2\phi - (d_{15} + d_{31}) \sin 2\phi] \sin\theta \cos\theta$ |
| $\bar{4}$2m | $-d_{36} \sin 2\phi \sin\theta$ | $(d_{14} + d_{36}) \cos 2\phi \sin\theta \cos\theta$ |

The calculation of effective nonlinear coupling constants for biaxial crystals is considerably more tedious although the basic approach remains the same.[36–41] Among the promising new nonlinear crystals, three (namely $KTiOPO_4$, $KNbO_3$ and $LiB_3O_5$) belong to the orthorhombic *mm2* point group and thus are optically biaxial. In light of the increasing importance of these materials in OPO applications, we briefly outline here the problem of evaluating the coupling constant, $d_{eff}$, in these materials. Our approach follows closely Hobden's classic 1967 paper on the subject.[42]

Analogous to the uniaxial crystal case, the effective coupling constant, $d_{eff}$, is obtained from the projection of the $\mathbf{P}(\omega)$ and $\mathbf{E}(\omega)$ fields onto the nonlinear susceptibility tensor. Since the phase matching requirement dictates the orientation of the polarization fields, we need to first work out the polarization eigenmodes of light propagating along arbitrary direction in a biaxial crystals.[43] The refractive indices of the two orthogonal polarizations in a biaxial crystal are given by the two real roots, $n_{e,1}$ and $n_{e,2}$, of the generalized Fresnel equation:

$$\frac{\sin^2\theta\cos^2\phi}{[n^{-2}-n_x^{-2}]}+\frac{\sin^2\theta\sin^2\phi}{[n^{-2}-n_y^{-2}]}+\frac{\cos^2\theta}{[n^{-2}-n_z^{-2}]}=0 \tag{5.5}$$

where $n_x, n_y$ and $n_z$ are the crystal refractive indices when the polarizations are along the $x$, $y$ and $z$ directions respectively. If the principal dielectric axes are chosen such that $n_z > n_y > n_x$, then the two real roots to Eq. (5.5) obey the relation $n_{e,1} > n_{e,2}$, and the corresponding polarization directions, $\mathbf{P}_{e,1}$ and $\mathbf{P}_{e,2}$, can be defined with the help of Figure 5.3. In a uniaxial crystal, the two polarization directions always lie in the constant $\theta$ or constant $\phi$ plane. In a biaxial crystal, this is no longer true, and the polarizations $\mathbf{P}_{e,1}$ and $\mathbf{P}_{e,2}$ tilt away from these planes by an amount $\delta$ which varies with the phase velocity propagation direction $\mathbf{k}$ specified by the angle $\{\theta, \phi\}$. In Figure 5.3, $2\Omega$ (often referred to as the *optic angle*) is the angle between the two optic axes lying in the $xz$ plane and is given by:

$$\tan^2\Omega=\frac{n_x^{-2}-n_y^{-2}}{n_y^{-2}-n_z^{-2}}. \tag{5.6}$$

As in the case of uniaxial crystal, it is convenient to classify the birefringence of a biaxial crystal as positive when $2\Omega < 90°$, and negative, $2\Omega > 90°$.

The polarization tilting angle $\delta$ is given by the expression:

$$\tan 2\delta=\frac{\cos\theta\sin 2\phi}{\cot^2\Omega\sin^2\theta-\cos^2\theta\cos^2\phi+\sin^2\phi} \tag{5.7}$$

which can be readily derived using vector algebra. Figure 5.4 plots the behavior of $\delta$ for different propagation direction $\{\theta, \phi\}$ in the material KTP, which has an optic angle of $2\Omega \sim 35°$. Near the two optic axes, the two allowed polarizations, $\mathbf{P}_{e,1}$

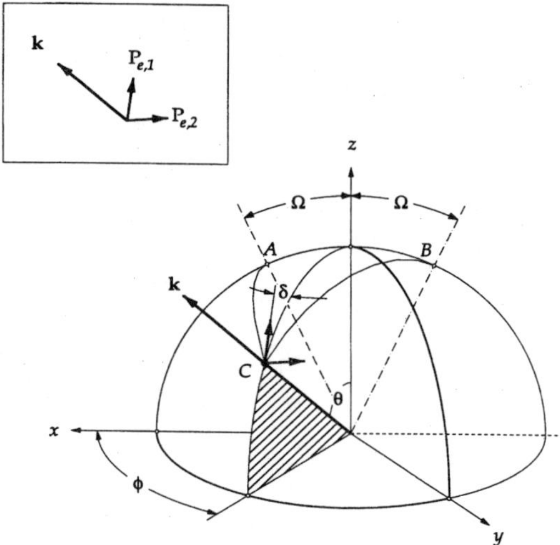


**Fig. 5.3** Polarization in an optically biaxial crystal with *optic angle* $2\Omega$. The $\mathbf{P}_{e,1}$ wave is parallel to the bisector of the spherical angle *ACB*, and is rotated away from the $\mathbf{k}z$ plane (shaded) by an amount $\delta$ given by Eq. 5.7. $\{x, y, z\}$ denote the principal dielectric axes with $n_z > n_y > n_x$. For clarity, the two polarization states $\mathbf{P}_{e,1}$ and $\mathbf{P}_{e,2}$ are defined in the inset.

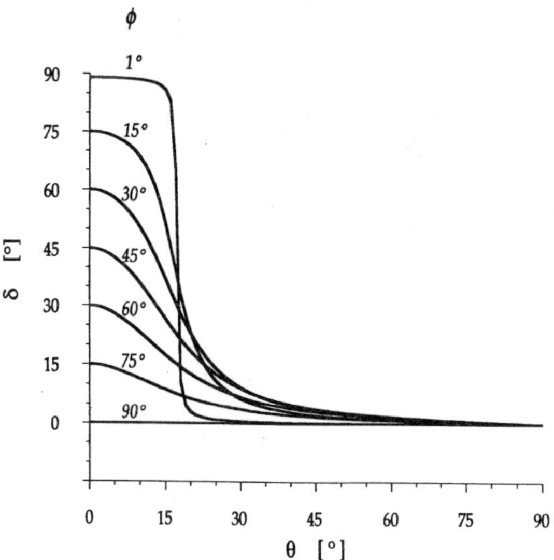


**Fig. 5.4** Polarization tilting angle ($\delta$) of $\mathbf{P}_{e,1}$ wave in biaxial $KTiOPO_4$ (optic angle, $2\Omega \sim 35°$). $\{\theta, \phi\}$ denotes light propagation direction $\mathbf{k}$ (see Figure 5.3). The $\mathbf{P}_{e,2}$ wave is orthogonal to $\mathbf{P}_{e,1}$.

and $\mathbf{P}_{e,2}$, sweep rapidly through 90°. Physically, this rapid change in $\delta$ corresponds to the discontinuity of the normal surfaces in a biaxial crystal. It is a manifestation of the fact that in the $xz$ plane, the refractive indices of the in-plane polarization and the out-of-plane polarization cross at half the optic angle, $\Omega$. Boulanger and Marnier have reported the experimental observation of $\delta$ using a KTP crystal sphere.[44] Except for the value of the optic angle, the behavior of $\delta$ is qualitatively similar for all biaxial crystals.

Since $n_{e,1} > n_{e,2}$ for all propagation directions, the collinear phase matching condition in a biaxial crystal can be expressed as:

$$\text{Type I} \qquad n_{e,2}(\omega_p) = \frac{\omega_s}{\omega_p} n_{e,1}(\omega_s) + \frac{\omega_i}{\omega_p} n_{e,1}(\omega_i), \tag{5.8}$$

$$\text{Type II} \qquad n_{e,2}(\omega_p) = \frac{\omega_s}{\omega_p} n_{e,2}(\omega_s) + \frac{\omega_i}{\omega_p} n_{e,1}(\omega_i).$$

Despite the simple appearance of Eq. (5.8), calculating the phase matching problem can be a good bit more complicated. In crystals with symmetry lower than orthorhombic (e.g. L-arginine phosphate, which is monoclinic),[45] material dispersion can in theory lead to a wavelength dependent rotation of the principal dielectric axes relative to the piezoelectric (i.e. crystallographic) axes and care must be taken to 'calibrate' this vectorial dispersion of the crystallographic axes with the principal dielectric axes. In these cases, using a spherical crystal[3,46] is perhaps the most efficient way to map out the linear and nonlinear optical properties of low symmetry materials. Indeed, by sum-mixing the signal and idler outputs of an OPO[47] in a spherical crystal, the energy conservation condition is automatically satisfied, and a complete set of parametric tuning curves can in principle be measured for a new material without any prior knowledge of the Sellmeier equations. For a material whose dispersion is negligible, the numerical evaluation of Eq. (5.8) can be readily performed starting with sufficiently accurate Sellmeier equations.[36,48]

The principal dielectric axes $\{x, y, z\}$, with $n_z > n_y > n_x$, are chosen above to simplify our analysis of the biaxial crystal optics problem. By convention,[32,49] the nonlinear susceptibility tensor is defined relative to the piezoelectric {X,Y,Z} axes, which in turn are defined by the crystallographic {a,b,c} axes and the direction of spontaneous polarization $P_s$. To evaluate the effective coupling constant, Hobden has proposed that the nonlinear susceptibility tensor be transformed into the principal dielectric axes.[42] With this transformation, the analytical evaluation of $d_{\text{eff}}$ becomes a straightforward, albeit still tedious, endeavor.[38,41]

For operation simplicity, nearly all OPO's reported to date using biaxial crystals are phase-matched in their principal dielectric planes. This simplifies the problem to a special case of uniaxial crystal optics. With $\mathbf{k}$ lying in a principal plane, the polarization normal to the plane coincides with the crystal tuning axis, and its refractive index no longer depends on the phase matching angle. Denoting this as the 'ordinary' wave, the effective coupling can be obtained following the procedure outlined for BBO above. Note that for phase matching in the plane containing the optic axes, the intersection of the two normal surfaces at $\Omega$ leads to a sign change in the crystal birefringence. To satisfy the phase matching condition, the role of the extraordinary and ordinary wave must be interchanged when the phase matching angle crosses $\Omega$. In the general case of arbitrary phase matching direction $\{\theta, \phi\}$ above, this sign change shows up as a rapid change in $\delta$ near the optic axes (see Figure 5.4).

Due to the increasing importance of *mm2* materials in nonlinear optics, we have summarized the expressions for the principal plane coupling constants of this point group in Table 5.2.[41] To be consistent with our discussion above and to allow easy comparison with literature, we have used the angles $\{\theta, \phi\}$ to refer to the optical $\{x, y, z\}$ axes as defined by Eq. 5.7 and in Figure 5.3. The contracted indices $\{l, m\}$ of the tensor element $d_{ij}$ normally denote the piezoelectric {X,Y,Z} axes. However, to be consistent with the convention already adopted for several commonly used crystals, $\{i, j\}$ in Table 5.2 refer to the crystallographic $\{\mathbf{a}, \mathbf{b}, \mathbf{c}\}$ axes instead. (See also Tables 5.4 and 5.5.)

**Table 5.2** Principal plane $d_{\mathrm{eff}}^{\mathrm{SHG}}(\theta, \phi)$ (expressions for crystals with *mm2* symmetry. $\{\theta, \phi\}$ and $\{x, y, z\}$ refer to the dielectric axes, while indices *ij* refer to the crystallographic $\{\mathbf{a}, \mathbf{b}, \mathbf{c}\}$ axes. Note that within each of the four regions below, either Type I or Type II phase-matching is efficient but not both.

| | | $xy$ | $yz$ | $xz;\theta>\Omega$ | $xz;\theta<\Omega$ |
|---|---|---|---|---|---|
| $n_c>n_b>n_a$ | I | 0 | 0 | 0 | $-d_{32}\sin\theta$ |
| e.g. $KTiOPO_4$ | II | $d_{24}\cos^2\phi+d_{15}\sin^2\phi$ | $d_{15}\sin\theta$ | $d_{24}\sin\theta$ | 0 |
| $n_c>n_a>n_b$ | I | 0 | 0 | 0 | $-d_{31}\sin\theta$ |
| | II | $d_{15}\cos^2\phi+d_{24}\sin^2\phi$ | $d_{24}\sin\theta$ | $d_{15}\sin\theta$ | 0 |
| $n_a>n_b>n_c$ | I | $d_{31}\sin\phi$ | $d_{32}\cos^2\theta+d_{31}\sin^2\theta$ | 0 | $d_{32}\cos\theta$ |
| | II | 0 | 0 | $-d_{24}\cos\theta$ | 0 |
| $n_b>n_a>n_c$ | I | $d_{32}\sin\phi$ | $d_{31}\cos^2\theta+d_{32}\sin^2\theta$ | 0 | $d_{31}\cos\theta$ |
| e.g. $KNbO_3$ | II | 0 | 0 | $-d_{15}\cos\theta$ | 0 |
| $n_b>n_c>n_a$ | I | $-d_{32}\cos\phi$ | 0 | $-d_{31}\cos^2\theta-d_{32}\sin^2\theta$ | 0 |
| e.g. $LiB_3O_5$ | II | 0 | $-d_{15}\cos\theta$ | 0 | $-d_{15}\cos^2\theta-$ $d_{24}\sin^2\theta$ |
| $n_a>n_c>n_b$ | I | $-d_{31}\cos\phi$ | 0 | $-d_{32}\cos^2\theta-d_{31}\sin^2\theta$ | 0 |
| | II | 0 | $-d_{24}\cos\theta$ | 0 | $-d_{24}\cos^2\theta-$ $d_{15}\sin^2\theta$ |

As mentioned earlier, beam-'walkoff' can limit the interaction length during nonlinear frequency conversion. Specifically, in an optical parametric oscillator, two effects, namely refraction and energy propagation, can lead to reduced spatial overlap of the pump and signal beams.[33] Consider first light refraction at the surface of a uniaxial crystal. For simplicity, we assume the optic axis of the crystal lies in the plane of incidence. This is the case for most commonly used phase-matching geometry. Conservation of the photon momentum component along the air-crystal interface gives the condition $\mathbf{k}_{i,x} = \mathbf{k}_{r,x}$ or $\sin i = n_e(\theta) \sin r$. Along with Eq. (2.6), Snell's Law for an anisotropic uniaxial crystal becomes:

$$\cot r = \frac{(e^2 - o^2)\sin 2\alpha + 2\sqrt{o \sin^2 \alpha / \sin^2 i + e \cos^2 \alpha / \sin^2 i - o^2 e^2}}{2(o^2 \sin^2 \alpha + e^2 \cos^2 \alpha)} \tag{5.9}$$

where $o = 1/k_o$, and $e = 1/k_e$. In Eq. (5.9), $i$ and $r$ are the angle of incident and refraction respectively, and $\alpha$ denotes the angle between the optic axis and the crystal surface (see Figure 5.5). Except for the special case of normal incident, refractive index dispersion leads to different refraction angles $r$, for the pump, signal and idler when the OPO is angle-tuned.

In addition, the anisotropy of the medium also renders the polarization vector $\mathbf{P}(\omega)$ noncollinear with the electric field $\mathbf{E}(\omega)$. As energy propagates along the direction of the Poynting vector $\mathbf{s} = \mathbf{E} \times \mathbf{H}$, the phase velocity vector $\mathbf{k}$, which is orthogonal to $\mathbf{P}$ and $\mathbf{H}$, becomes noncollinear with the direction of energy propagation, $\mathbf{s}$. The angular separation between $\mathbf{s}$ and $\mathbf{k}$ (i.e. $\mathbf{P}$ and $\mathbf{E}$) is known as the walkoff angle $\rho = (\theta_s - \theta)$ where $\theta_s$ is the angle between the optic axis and $\mathbf{s}$ given by the expression:

$$\tan \theta_S = \left(\frac{n_o}{n_e}\right)^2 \tan \theta \tag{5.10}$$

In a non-magnetic crystal, $\rho$ lies in the plane containing $\mathbf{P}$, $\mathbf{E}$, $\mathbf{k}$ and $\mathbf{s}$. In Eq. (5.10), $n_o$ and $n_e$ are the refractive indices between which the extraordinary wave index $n_e(\theta)$ can vary. For Type I phase matching where the pump and the signal/idler have orthogonal polarizations, $\rho$ becomes the angle with which the pump separates from the signal/idler as they propagate through the crystal. For Type II phase-matching in a negatively uniaxial crystal (e.g. $\beta$-$BaB_2O_4$), one of the two parametric output waves has the same polarization with the extraordinary pump wave. The angular separation between the Poynting vectors of these two extraordinary waves now depends on the *dispersion* of $\rho$, i.e. $\rho(\omega_p) - \rho(\omega_{s,i})$, which is significantly smaller than $\rho$ in most crystals. By selecting this output wave as the resonating signal in an OPO, beam walkoff problem can be reduced, and longer crystal length can be used to increase the nonlinear gain.

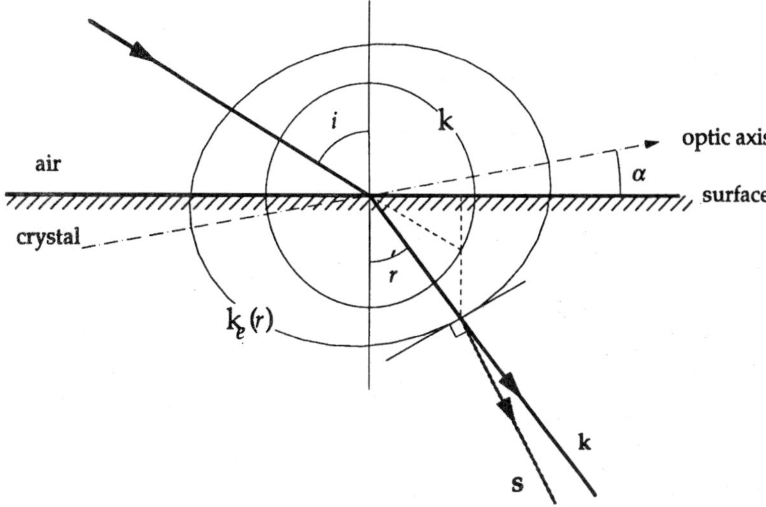


**Fig. 5.5** Snell's law for light refraction into an anisotropic medium.[33] The circle and the ellipse are the **k**-vector surfaces for the incident wave in air and in the crystal (depicted as an extraordinary wave) respectively. Conservation of photon momentum along air-crystal interface requires that $\mathbf{k}_{i,x} = \mathbf{k}_{r,x}$ as shown by the dotted lines.

## 5.2 MATERIAL SURVEY

During the past decade, a great deal of effort has been devoted to the development of organic molecular crystals for nonlinear optics. Systematic investigations have led to an empirical understanding of the molecular nonlinearity,[1,50] and crystals with phenomenally large optical nonlinearity have been reported.[51–55]

These crystals unfortunately have only limited applications in OPO's due to their relatively narrow transparency range. Their short wavelength absorption cutoffs are generally limited to ~0.4–0.5 $\mu$m by the $\pi-\pi^*$ transition of the extended conjugation, while overtone C-H absorptions often become a problem in the near infrared ~1.0–1.2 $\mu$m.[51] This narrow optical transparency also makes these crystals strongly dispersive, and a large crystal birefringence ($\Delta n \sim 0.2$–$0.4$) is often needed to achieve phase-matching. Although large crystal birefringence is common among organic molecular crystals, it leads to serious walk-off and angular acceptance problems which must be overcome before efficient oscillation can be obtained.[51]

Experimental OPO's in the earlier days tend to be in the infrared, and the widely used crystal $LiNbO_3$ offers a potential output tuning range between 0.8–3.5 $\mu$m.

Extension to the shorter wavelength part of the spectrum had been more difficult due to the lack of efficient nonlinear optical materials that can be phase matched in the UV. The first nonlinear material to break this wavelength barrier was crystalline urea.[60] A continuous tuning range from 500 nm to 1.2 $\mu$m was obtained when pumped at the third harmonic of a Nd:YAG laser. However, although crystalline urea has a very high single-shot damage threshold, it suffers from long-term, multi-shot optical damage problems[17] and was soon superseded by the new nonlinear optical crystals $\beta$-$BaB_2O_4$, which is comparatively much easier to grow and has superior crystal properties. These material characteristics have made it the preferred crystal for use in commercial OPO's operating in the visible and near UV.

More recently, the related lithium triborate (LBO) crystal has also appeared on the scene and is attracting considerable attention as a potentially useful OPO material. For the UV and visible range, this material is now clearly one of the materials of choice for high power OPO applications. From the deep red to about 3.0 $\mu$m, KTP can meet many of the key requirements for OPO applications and shows great promise. The development of the femtosecond and picosecond OPOs has shown rapid progress recently and promises to have a significant impact on the studies of a wide range of ultrafast phenomena. While KTP are currently the material of choice in these applications, recent progress in the material development of crystals such as $KNbO_3$, $KTiOAsO_4$, $CsTiOAsO_4$, and $RbTiOAsO_4$ are particularly promising as they offer design flexibility for optimizing specific output characteristics of these OPOs. The properties of these promising new nonlinear crystals will be discussed in more details below.

## 5.3 MID INFRARED OPO MATERIALS

In the mid IR region (2–20 $\mu$m), nonlinear optical materials suitable for optical parametric oscillator applications include the chalcopyrites ($AgGaSe_2$, $AgGaS_2$, $ZnGeP_2$ and $CdGaAs_2$), $Tl_3AsSe_3$, and GaSe. During the past decade, advances in infrared OPO technology are primarily driven by progress in the crystal growth of these materials and by the development of efficient diode-pumped $\sim$2 and $\sim$ 3 $\mu$m lasers. Early OPO experiments using the chalcopyrites were limited by small crystal size and poor optical quality. The origin of optical loss in as-grown chalcopyrite crystals was traced to microscopic scattering centers caused by compositional nonstochiometry during crystal growth. Through careful control of the growth conditions and post-growth annealing, crystals with low optical loss ($<0.016$ cm$^{-1}$ at 2.1 $\mu$m in $AgGaSe_2$) can now be commercially produced.[61] The attainable crystal size was also limited by crystal crackings caused by the anisotropic thermal expansion[62] of the crystal relative to the quartz ampoule used during Bridgman growth. Proper adjustment of the crystal growth parameters, such as better alignment of the crystal growth axis[63] or the use of the horizontal gradient freeze technique, have eliminated this cracking problem, and led to ~3 cm long crystals suitable for OPO applications. Recent progress in the crystal growth of these materials can be found in the excellent review by Feigelson and Route.[4]

With better quality crystals comes a better understanding of the laser damage characteristics of these materials. A survey of the reported optical damage threshold for $ZnGeP_2$, $AgGaS_2$ and $AgGaSe_2$ clearly shows that it is the energy fluence, not the peak intensity, that is responsible for optical damage in these materials. This suggests that optical damage in these materials is caused by absorption induced heating, similar to those observed in semiconductors.[13] Thus, despite the low damage threshold of these materials in the nanosecond regime, very high peak intensity[64] picosecond or femtosecond pulses can be used, opening the door for ultrafast spectroscopy in the 3–20 $\mu$m range. More importantly, it is found that surface optical damage thresholds for these materials are nearly one order of magnitude lower than their bulk thresholds, and depend strongly on their crystal surface conditions. While this has currently slowed progress in mid-infrared OPO technology, it offers the opportunity that, with proper surface coatings, significant improvement in device performance may be achievable.

The chalcopyrites have been the primary materials for infrared OPO applications. The selection between $ZnGeP_2$ and $AgGaSe_2$ amounts to a trade-off between optical transparency and optical nonlinearity. Table 5.3 lists the relevant parameters of existing infrared materials.[4,5,64–70] Based on published optical transmission spectra and OPO tuning curves, efficient tunable OPO outputs can be expected between ~3.2 to 8 $\mu$m with $ZnGeP_2$ and ~2.5 to 12 $\mu$m, $AgGaSe_2$. The reported optical nonlinearity, in mks units (with $d_{ij}[m/V] = \frac{4\pi}{3} \times 10^{-4} d_{ij}[esu]$) of $ZnGeP_2$ (~88 pm/V) is nearly three times larger than $AgGaSe_2$ (~32.4 pm/V), and $ZnGeP_2$ OPO with slope efficiencies as high as 37% have been reported.[71] Its optical absorption (~0.25 $cm^{-1}$) is unfortunately very high near the 2.1 $\mu$m pump wavelength. This exerts a ~50% toll in the available pump power, and significantly limits its overall energy efficiency. Worst yet, the strong absorption causes severe thermal problems which must be carefully managed if stable, efficient OPO operation is to be maintained.[71] In contrast, $AgGaSe_2$ with a minimal absorption coefficient of ~0.01 $cm^{-1}$ at 2.1 $\mu$m have been grown. Improvement in the parametric gain is thus directly related to the availability of longer crystals. While it is unclear if its 2.1 $\mu$m absorption can be further reduced, $ZnGeP_2$ remains a highly efficient material particularly when pumped at longer wavelengths.

The new nonlinear crystal $Tl_3AsSe_3$, with a nonlinearity of ~40 pm/V and a transparency similar to that of the selenogallate, may be an interesting alternative to $AgGaSe_2$ and $ZnGeP_2$. Like the chalcopyrites, the progress in the crystal growth of high quality $Tl_3AsSe_3$ is a classic example of the beneficial interplay between crystal defect characterization and single crystal preparation. An overview of the history of $Tl_3AsSe_3$ development can be found in references 72 and 67. Though little is known about the nonlinear optical properties of $Tl_3AsSe_3$, initial studies suggest that it should be an attractive OPO material. Using a 4.5 cm long crystal, conversion efficiency of 57% has been observed in the doubling of the 10.6 $\mu$m $CO_2$ laser: This is the highest SHG efficiency ever reported in the mid-IR.[27] Equally interesting is the reported 10 J/cm$^2$ (at 10.6 $\mu$m) surface damage fluence threshold,[27] which more than triples that of $AgGaSe_2$.[26] The crystal's birefringence is sufficiently high for phase-matching throughout its transparency range. Figure 5.6 shows the calculated[66] Type I and Type II OPO tuning curves for a 2.1 $\mu$m pumped OPO. The viability of these OPO configurations will depend on the crystal's absorption at 2.1 $\mu$m and its nonlinear susceptibility $d_{22}$ and $d_{31}$.

In Table 5.3, we have listed the material parameters traditionally used for comparing nonlinear crystals suitable for OPO applications.[13] These parameters, including the nonlinear figure-of-merit $d_{\mathrm{eff}^2}/n_p n_s n_i$, optical damage threshold $G_{\mathrm{burn}}$ and optical transparency, remain as valid now as when they were first proposed. During the past decade, as OPO technology matures from mere laboratory curiosity towards commercial application, more emphasis is being placed on factors such as ease of operation, reliability and cost effectiveness. These pragmatic considerations favor materials with the best overall material characteristics, including its mechanical strength, chemical inertness, temperature stability, compatibility with existing coating and polishing technology, and the ease of crystal growth. The importance of these factors have recently been reiterated in the excellent review by Eimerl *et al.*[56] Thus, information presented in these tables are meant to be a 'guideline' comparison only, and other factors not readily quantifiable must be taken into account in actual device design.

## 5.4 VISIBLE AND NEAR-INFRARED OPO MATERIALS

Early OPO's operating in this wavelength region were primarily pumped by the fundamental and second harmonics of the Nd:YAG. In these devices, $LiNbO_3$ and $LiIO_3$ were the most common nonlinear crystals due to their availability in large size and high quality. Although these early OPO's could be operated reliably with proper design,[73] the low damage thresholds of $LiNbO_3$ and $LiIO_3$ crystals required the use of specialized pump lasers with highly uniform intensity profiles. Various material property limitations of $LiNbO_3$ and $LiIO_3$ also made important device characteristics, such as efficient output coupling, pulse length, spectral linewidth...etc, difficult to be incorporated. During the past decade, the development of high quality crystals of $KNbO_3$, $KTiOPO_4$ and its isomorphs, and various damage resistance lithium niobates promises to overcome some of these difficulties, providing continuous wavelength tunability between 0.8–4 $\mu$m. In this section, the nonlinear optical properties of these materials are reviewed.

**Table 5.3** Properties of selected nonlinear crystals for mid infrared OPOs.

| Crystals | $Tl_3AsSe_3$ | $AgGaSe_2$ | $ZnGeP_2$ | GaSe |
|---|---|---|---|---|
| point group | 3m | $\bar{4}2m$ | $\bar{4}2m$ | $\bar{6}2m$ |
| cell constants [Å] | **a** = 9.80<br>**c** = 7.08 | **a** = 5.988<br>**c** = 10.884 | **a** = 5.7578<br>**c** = 10.3072 | **a** = 3.749<br>**c** = 15.907 |
| transparency [$\mu$m] | 1.3–17 | 0.73–17 | 0.74–12 | 0.65–18 |
| 2.05 $\mu$m pumped, Type I OPO polarization | $e{-}{>}o{+}o$ | $e{-}{>}o{+}o$ | $o{-}{>}e{+}e$ | $e{-}{>}o{+}o$ |
| tuning range [$\mu$m] | 2.3–17 | 2.5–12 | 2.6–10 | 3.5–18[(b)] |
| refractive indices $n_e$ | 3.227 | *2.6060*[(a)] | 3.1889 | *2.488* |
| @ ~2 $\mu$m $n_o$ | 3.419 | *2.6365* | 3.1490 | *2.854* |
| $\theta_{\text{pm}}$ range [°] | *23–31* | 43–50 | 50–55 | 9–13 |
| walkoff, $\rho$ [°] | 2.8 | 0.7 | -0.7 | 3.3 |
| crystal sizes[(c)] [cm] | 4 | 3 | 1 | 1 |
| nonlinear figure of merit $d^2/n^3$ [pm/V]$^2$ | 45 | 57 | 187 | 180 |
| $\alpha$ @ 2.1 $\mu$m [cm$^{-1}$] | n.a. | <0.01 | 0.26 | n.a. |
| surface damage threshold [J/cm$^2$] | 10[(d)] | 3 | 0.5 | >3.3 |
| Sellmeier Equation | $n_i^2(\lambda)=A+\frac{B}{1-(C/\lambda)^2}+\frac{D}{1-(E/\lambda)^2}$ | | | $(e)$ |

| | $n_e$ | $n_o$ | $n_e$ | $n_o$ | $n_e$ | $n_o$ | $n_e$ | $n_o$ |
|---|---|---|---|---|---|---|---|---|
| $A$ | 1 | 1 | 3.3132 | 3.9362 | 4.63318 | 4.4733 | 6.06 | 8.038 |
| $B$ | 8.993 | 10.210 | 3.3616 | 2.9113 | 5.34215 | 5.26576 | 0.5754 | 0.526 |
| $C[\mu m]$ | 0.444 | 0.444 | 0.38201 | 0.38821 | 0.37756 | 0.3658 | 0.21284 | 0.06 |
| $D$ | 0.308 | 0.522 | 1.7677 | 1.7954 | 1.45795 | 1.49085 | | 0.00082 |
| $E[\mu m]$ | 25 | 25 | 40 | 40 | 25.74 | 25.74 | 0.03225 | $2.7\times10^{-6}$ |

(a) calculated values are shown in italics.

(b) GaSe tuning range shown is with a 2.9 $\mu$m pump source.

(c) crystal length is along phase-matching direction. All crystals are grown from variation of the Bridgman technique.

(d) measured at 10.6 $\mu$m.

(e) Sellmeier equations for GaSe are:
$n_e^2(\lambda)=A+B/(\lambda^2-C^2)-(E\lambda)^2$; and
$n_o^2(\lambda)=A+B/\lambda^2-C/\lambda^4-D\lambda^2-E\lambda^4$, where units of $B,C,D,E$ are such that $n(\lambda)$ is unitless.

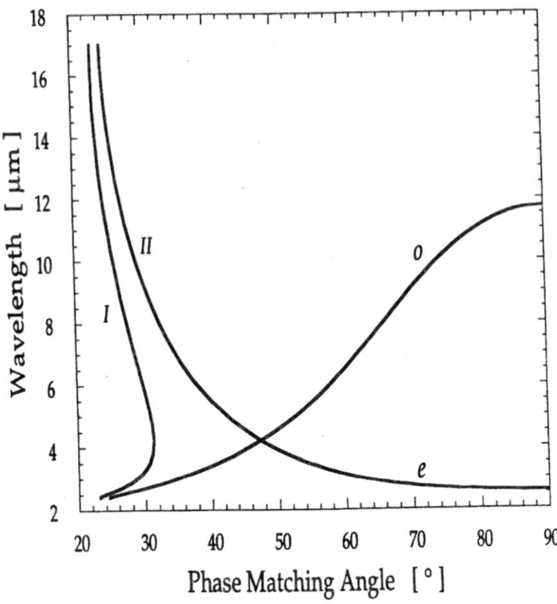


**Fig. 5.6** Calculated Type I and II OPO tuning curves for $Tl_3AsSe_3$. Pump wavelength is 2.1 $\mu$m, and the Sellmeier equations are taken from Ewbank *et. al.*[66]

Fueled by its use as surface acoustic wave filters in consumer electronics, the crystal growth technology for $LiNbO_3$ has reached a level of sophistication[74] unmatched by other nonlinear crystals except perhaps $KH_2PO_4$ (KDP). Unfortunately, undoped $LiNbO_3$ suffers photorefractive damage, making it unsuitable for many applications at wavelengths shorter than 1 $\mu$m. Although, the MgO:$LiNbO_3$ composition have been grown in 1970,[75] its increased photorefractive damage resistance was not recognized until 1980 by Zhong and coworkers.[76] Since then, other effective dopants, such as ZnO[77] and $Sc_2O_3$,[78] have also been identified. Except for minor modifications in crystal birefringence and dispersion, the nonlinear optical properties of these doped crystals remain unchanged and readers interested in the tuning characteristics of $LiNbO_3$ OPO should refer to published literature.[13] Preliminary optical damage studies suggested that with MgO doping, $LiNbO_3$'s intrinsic dielectric breakdown threshold ($\sim$14 J/cm$^2$ at 1.053 $\mu$m) may be realizable.[79] This should significantly increase the versatility of $LiNbO_3$ for visible and near IR applications. However, crucial material related problems, such as composition homogeneity, scattering centers and surface damage, remain and further advances in $LiNbO_3$-based OPO technology will depend on the timely resolution of these problems.[28–30]

### 5.4.1 Potassium Titanyl Phosphate (KTP) and Isomorphs

KTP is the best known member of the isomorphic family of nonlinear optical crystals with the generic chemical formula: $M\text{TiO}XO_4$, where $X$ = {P and As}, $M$ = {$NH_4$, K, Rb, Tl and Cs (for $X$=As only)}. The synthesis of the titanyl phosphate isomorphs was first reported by Masse and Grenier in 1971.[80] The attractive nonlinear optical properties of these crystals and their arsenate derivatives were recognized by researchers at DuPont in the early 1970's during a systematic survey of novel optical materials.[81] Unfortunately, KTP was discovered at a time when interests in inorganic frequency conversion materials reached a lull. Persistent military interests in this material, however, have led to the successful development of viable KTP crystal growth technologies,[82,83] which are partly responsible for the renewed interests in OPO technology.

The KTP family of crystals is one of the most versatile nonlinear optical materials for frequency conversion applications.[84] Besides having excellent properties (see Table 5.4) traditionally valued in a nonlinear crystal, such as large temperature bandwidth, high damage threshold, broad transparency and low optical absorption, these isomorphs offer other interesting properties which give them unique advantages in practical applications. For instance, the cation exchange property of these crystals[85] makes them particularly suitable for micro-optic applications such as waveguide SHG, monolithic electro-optic modulators, and all optical switching. In addition, recent experiments indicate that these crystals are useful for piezoelectric and acousto-optic devices as well.[86,87] Such broad based interest is helpful in sustaining the costly development of the technology needed to grow large, high quality single crystals suitable for use in OPO's.

KTP belongs to the point group *mm2* (space group $Pna2_1$), with orthorhombic lattice constants **a**=12.814 Å, **b**=6.404 Å and **c**=10.616 Å. It has five nonvanishing second order susceptibility tensor elements {$d_{15}$, $d_{31}$, $d_{24}$, $d_{32}$ and $d_{33}$}, the magnitudes of which[88] are summarized in Table 5.4. The nonlinearity of KTP results mainly from the two different sets (namely *cis*- and *trans*-) of distorted $TiO_6$ octahedra in the crystals. The short-long O=Ti–O bonds, which define the direction of the distortion, of the *cis*- and *trans*- octahedra point approximately along the [011] and [201] directions respectively. As the cell constants of KTP have $|\mathbf{a}| \sim 2|\mathbf{b}|$, the [011] and [201] directions lie in the $yz$ and $xz$ planes respectively, and have similar polar angles measuring from the **c**-axis. This 'pseudo-*4mm*' symmetry of the O=Ti–O bonds has been used to explain the comparable magnitudes of $d_{15}$ and $d_{24}$ in KTP.[89] Figure 5.7 shows the complete crystal structure of KTP viewed along the polar **c**-axis. It also shows the *4mm* pseudosymmetry of the short Ti=O bonds.

**Table 5.4** Properties of selected nonlinear crystals for visible and near infrared OPO's.

| Crystals | $LiNbO_3^{(a)}$ | $KTiOPO_4$ | $KNbO_3$ |
|---|---|---|---|
| point group | 3m | *mm2* | *mm2* |
| cell constants | **a** = 5.1494 Å<br>**c** = 13.862 Å | **a** = 12.814 Å<br>**b** = 6.404 Å<br>**c** = 10.616 Å | **a** = 5.6896 Å<br>**b** = 3.9692 Å<br>**c** = 5.7256 Å |
| axes assignment<br>crystallographic | **a** **c** | **a** **b** **c** | **a** **b** **c** |
| piezoelectric | X⊥*m* Z | X Y Z | X Y Z |
| dielectric/optical | *x* *z* | *x* *y* *z* | *x* *y* *z* |
| transparency [$\mu$m] | 0.32–4.5 | 0.35–4.3 | 0.39–5.5 |
| efficient phase matching configuration | $e{-}{>}o{+}o$ | $o{-}{>}e{+}o(xz)$ | $e{-}{>}o{+}o(xz)$ |
| tuning angle | temp./angle | angle | temp./angle |
| refractive indices<br>@ $\sim$1.064 $\mu$m | $n_o$=2.2325<br>$n_e$=2.1560 | $n_{\mathbf{c}}$=1.8304<br>$n_{\mathbf{b}}$=1.7469<br>$n_{\mathbf{a}}$=1.7400 | $n_{\mathbf{b}}$=2.2576<br>$n_{\mathbf{a}}$=2.2195<br>$n_{\mathbf{c}}$=2.1194 |
| nonlinearity$^{(b)}$ $d_{ij}$<br>[pm/V] | $d_{31}$=4.8<br>$d_{22}$=2.3<br>$d_{33}$=29.7 | $d_{15}$=1.9<br>$d_{24}$=3.6<br>$d_{31}$=2.5<br>$d_{32}$=4.4<br>$d_{33}$=16.9 | $d_{15}$=16.5<br>$d_{24}$=17.1<br>$d_{31}$=15.8<br>$d_{32}$=18.3<br>$d_{33}$=27.4 |
| $\Delta T.\ell$[°C-cm] | ~0.3 | 10–300 | 0.1–0.3 |
| crystal damage threshold<br>@ 1.064 $\mu$m [GW/cm$^2$] | 1–14$^{(c)}$ | >1.2 | 2$^{(d)}$ |
| phase transition temperature [°C] | ~1100 | ~940 | 225 & 435 |
| crystal growth | Czochralski | flux or hydrothermal | flux |
| growth limitation | composition uniformity | slow growth rate | slow growth rate |
| post growth processing | poling | none | poling & detwinning |
| crystal sizes [cm] | 10 | 2 | 2 |

(a) values for congruent melting $LiNbO_3$.
(b) $ij$ refer to piezoelectric (X, Y, Z) axes. Note that $d_{ij}(m/V) = \frac{4\pi}{3} \times 10^{-4} d_{ij}(esu)$.
(c) values for MgO:$LiNbO_3$ (see Ref. 79).
(d) value for hydrothermal $KTiOPO_4$, which is higher than for flux gown material.

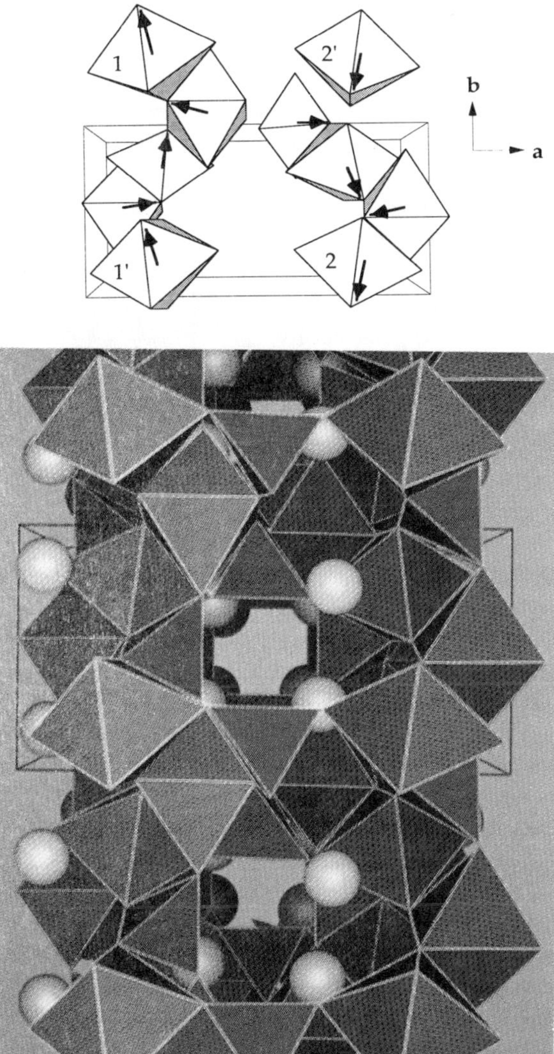


**Fig. 5.7** Crystal structure of $KTiOPO_4$ viewed along the **c**-axis. Top drawing shows the two types of helical $TiO_6$ octahedral chains (labelled as 1 and 2) which run parallel to the $[01\bar{1}]$ and $[0\overline{11}]$ directions. Arrows mark the **ab** plane projection of the short Ti=O bonds to illustrate the *4mm* "pseudo-symmetry" which makes $d_{15}$ comparable to $d_{24}$ (see text). $(1,1')$ denotes equivalent octahedra within adjacent $TiO_6$ chains of the same type. Same for $(2,2')$.

Theoretical calculations aimed at understanding the microscopic origin of the nonlinearity in KTP have met only limited success. More convincing evidence linking the optical nonlinearity to the $TiO_6$ octahedra came from direct powder SHG measurement of the numerous chemical isomorphs of KTP. Whereas the optical nonlinearity remains unchanged for the chemical replacement of $M$ and/or $X$, it decreases drastically whenever the titanyl group is replaced by groups with less bond distortion.[90]

The optical birefringence of KTP obeys the relation $n_{\mathbf{c}} > n_{\mathbf{b}} > n_{\mathbf{a}}$ such that the principal dielectric axes are related to the piezoelectric and crystallographic axes by $\{x = \mathrm{X} = \mathbf{a}\}$, $\{y = \mathrm{Y} = \mathbf{b}\}$ and $\{z = \mathrm{Z} = \mathbf{c}\}$. In this case, assuming that Kleinman symmetry is valid {i.e. $d_{15} \sim d_{31}$ and $d_{24} \sim d_{32}$}, the $d_{\mathrm{eff}}$ for SHG is given by

$$\begin{aligned} d_{\mathrm{eff}}(\mathrm{I}) =&\{(d_{24} - d_{15}) \sin 2\phi \cos\theta (3\sin^2\delta - 1)\cos\delta \\ &+ 3(d_{15}\cos^2\phi + d_{24}\sin^2\phi)\cos^2\theta\cos^2\delta\sin\delta \\ &+ (d_{15}\sin^2\phi + d_{24}\cos^2\phi)(3\sin^2\delta - 2)\sin\delta \\ &+ d_{33}\sin^2\theta\cos^2\delta\sin\delta\}\sin\theta \end{aligned} \tag{5.12}$$

for Type I; and by

$$\begin{aligned} d_{\mathrm{eff}}(\mathrm{II}) =& -\{(d_{15} - d_{24}) \sin 2\phi \cos\theta (3\cos^2\delta - 1)\sin\delta \\ &+ 3(d_{15}\cos^2\phi + d_{24}\sin^2\phi)\cos^2\theta\sin^2\delta\cos\delta \\ &+ (d_{15}\sin^2\phi + d_{24}\cos^2\phi)(3\cos^2\delta - 2)\cos\delta \\ &+ d_{33}\sin^2\theta\sin^2\delta\cos\delta\}\sin\theta \end{aligned} \tag{5.13}$$

for Type II. Far away from the optic axes, i.e. $\theta > 40°$, $\delta$ is small (see Figure 5.4), and $\sin\delta$ can be set to zero leaving:

$$d_{\mathrm{eff}}(\mathrm{I}) \approx -\{(d_{24} - d_{15})\sin 2\phi\}\cos\theta\sin\theta \tag{5.14}$$

and

$$d_{\mathrm{eff}}(\mathrm{II}) \approx -\{d_{15}\sin^2\phi + d_{24}\cos^2\phi\}\sin\theta \tag{5.15}$$

Several points can be made here. First, the $\sin 2\phi \sin 2\theta$ dependence of Eq. (5.14) above means that no Type I interaction can be effective in the principal planes. Second, unless $d_{33}$ is large (i.e. $d_{33}/d_{24} \sim 10$), the contribution of $d_{33}$ to $d_{\text{eff}}$(II) will be small due to the $\sin^3\theta \sin^2\delta \cos\delta$ variation in the last term of Eq. (5.13). Third, the effective coupling constant depends very much on the relative sign and magnitude of $d_{24}$ and $d_{15}$. As pointed out earlier, $d_{24}$ and $d_{15}$ in KTP are comparable due to the pseudo-*4mm* symmetry of the long-short O=Ti–O bond pairs. Thus, according to Equations (5.14) and (5.15), depending on their relative signs, efficient Type I phase matching may not be realizable. Direct measurement of Type I SHG in KTP reveals that $d_{24}$ and $d_{15}$ have the same sign.[91] This limits KTP's usefulness only to Type II phase-matching, which is ~50% less effective than Type I in using the available crystal birefringence. Thus despite KTP's large birefringence ($n_z - n_x \sim 0.1$), continuous extended wavelength tuning cannot be obtained from KTP-based OPO's when the pump is shorter than ~500 nm.

The usefulness of our approximate $d_{\text{eff}}$ expressions stems from the small optic angle ($2\Omega$) of KTP. This in turn is due to the quasi-uniaxial character (i.e. $n_z > n_y \gtrsim n_x$) of the crystal birefringence. The importance of this property for phase-matching has been discussed succinctly by Eimerl *et al.*[56] In a uniaxial crystal, the rotational symmetry of the index ellipsoid about the optic axis gives an angular bandwidth of $\Delta\phi\ell \sim 2\pi\ell$ in the $xy$ plane. In a biaxial crystal like KTP, this infinitely large angular bandwidth is traded-off for a small wavelength tuning range in the $\phi$ direction along with a large, but finite $\Delta\phi\ell$. Thus rotations in $\theta$ and $\phi$ in effect correspond to coarse and fine adjustments of the tuning rate. It is in fact this additional flexibility, coupled with the large temperature bandwidth ($\Delta T\ell \sim 20 - 170$°C-cm),[92] which make KTP such a versatile frequency doubler for the Nd-based lasers operating near 1 $\mu$m.

Efficient Type II phase matching has several inherent advantages over Type I processes. In Type II interaction, one of the two parametric output waves has the same polarization as that of the pump. By selecting this wave as the signal, i.e. as the resonating wave, the Poynting vector walkoff problem can be minimized. Also, since the signal and the idler outputs have different polarization in Type II phase matching, these waves are never truly degenerate. This leads to the significantly smaller tuning rate of Type II OPO's operating near the degenerate output wavelengths. As the main contribution of the spectral bandwidth of an OPO is from pump beam divergence, Type II OPO tends to show significantly better spectral characteristics near the degenerate point.

The nondegeneracy proves to be particularly important in the design of doubly resonant parametric oscillators (DRO) for frequency standard applications.[93] In a 2:1 frequency divider, it has been shown that Type II phase matching offers significantly better frequency stability than Type I since it effectively suppresses unwanted cluster modes.[94] Using a combination of temperature, angle and electro-optic tuning to phase-lock the signal and idler waves, Wong *et al.* have reported stable cw operation of KTP OPO's with a signal and idler beat-note linewidth of less than 25 mHz.[94,95] As the condition for phase matching is more restrictive than for interferometry,[33]

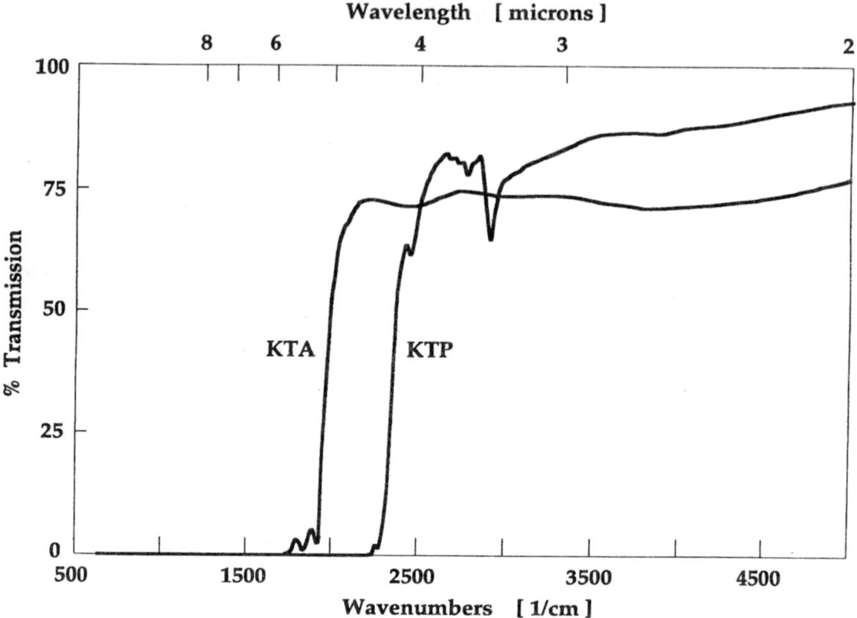


**Fig. 5.8** Optical transmission characteristics of $KTiOPO_4$ and $KTiOAsO_4$ in the near-infrared. The spectra were taken with unpolarized light using ~5 mm long and uncoated crystals. The 3.5 $\mu$m $PO_4$-group overtones prominent in KTP are shifted to ~ 4.2$\mu$m due to the As for P substitution. The overtone absorptions are also comparatively weaker in KTA. No significant difference in the UV cutoffs (at ~350 nm) is measured for the two materials.

these phase-locked cw OPOs promise frequency resolution several orders of magnitude better than the $10^{-10}$ currently available with interferometry.[93] Besides being useful for precision optical measurement, these devices also hold great promises in coherent optical communication applications.

Until recently, research on the KTP family of crystals has been focused primarily on the orthophosphate analogs. Recent advances have led to the crystal growth of large single domain crystals of the arsenate isomorphs of KTP suitable for OPO applications.[47,96,97] As pointed out earlier, orthophosphate overtone absorption near 3.5 $\mu$m limits the usefulness of KTP for parametric generation of 3–4 $\mu$m radiation. This overtone absorptions can be eliminated by replacing the phosphate with the heavier orthoarsenate $(AsO_4)^{-3}$ group.[98] Figure 5.8 illustrates the ~ 1 $\mu$m red-shift of the infrared cutoff of $KTiOAsO_4$ (KTA). The infrared cutoffs are similar for all the arsenate isomorphs, including $RbTiOAsO_4$ and $CsTiOAsO_4$. Thus, given their high optical damage thresholds, these arsenate crystals are promising alternative to the chalcopyrites for generating broadly tunable coherent radiation in the 3–5 $\mu$m region.

Growing concerns in environmental pollution has led to a need for reliable 'eyesafe' lasers suitable for remote sensing, ranging and process monitoring applications. Lasers emission near the 1.5 $\mu$m region is termed 'eye-safe' because optical absorption by occular fluid reaches a maximum near 1.54 $\mu$m, and thus lowers the risk of laser-induced retinal damage to the eye.[99] KTP enjoys a very special advantage in this application: pumped at 1.064 $\mu$m, an $x$-cut KTP OPO gives a noncritically phase-matched signal wavelength at 1.57 $\mu$m. Although the idler emission at 3.29 $\mu$m is strongly absorbed by KTP, conversion efficiency as high as 70% has been observed thanks to the broad temperature bandwidth of KTP.[100] However, for very high average power devices, this absorption can in principle still cause thermal dephasing problems, and in the worse scenario may lead to catastrophic crystal failure.

Fortunately, the recent successful development of single domain KTA crystals[96] has essentially eliminated this thermal-loading concern. Replacing KTP with KTA in the 'eye-safe' OPO shifts the noncritically phase-matched (NCPM) signal wavelength to 1.52 $\mu$m and recovers the 3.55 $\mu$m infrared idler as useful output.[47] As discussed earlier, the phosphate to arsenate substitution in KTA does not affect the optical nonlinearity. The substitution does however lead to changes in the linear optical properties, most important of which are the crystal birefringence and infrared absorption. Table 5.5 summarizes the linear and nonlinear optical properties of the KTP isomorphs.[101]

For the arsenate isomorphs, a systematic reduction in the crystal birefringence was observed with increasing formula mass.[101] As the crystal birefringence decreases, the phase matching angle for any pump and signal frequency combination shifts toward the $xy$ plane. Figure 5.9 shows the calculated 1.064$\mu$m pumped OPO tuning curves for KTA and CTA to illustrating this shift. The availability of a set of isomorphic crystals of KTP offers significant flexibility in device design. For instance, as shown in Figure 5.9, due to the $\sin\theta$ dependence of the nonlinear coupling, CTA will be a more efficient material than KTA for OPOs operating at the 2.128$\mu$m degeneracy, and for generating wavelengths longer than 3.5$\mu$m. More interestingly, by forming solid-solutions of these isomorphs, it may be possible to 'tune' the crystal birefringence to yield non-critically phase-matched crystal for operation within broad wavelength range. As in the case of MgO:$LiNbO_3$, the viability of this last approach will depend critically on the ability to grow solid solution of large ($\sim 5 \times 5 \times 15\text{mm}^3$) crystals free of compositional inhomogeneity.

KTP crystal growth technology has also experienced significant advances during the past decade. Since KTP melts incongruently at ~1148°C, crystals of KTP must be obtained by solution crystallization. Two growth methods are currently in commercial use: 1) high temperature flux growth and 2) hydrothermal crystal growth.

**Table 5.5** Linear and nonlinear optical properties of KTP isomorphs.

| Crystal | KTP | RTP | KTA | RTA | CTA |
|---|---|---|---|---|---|
| | Nonlinear susceptibilities at 1.064 $\mu$m [pm/V] $\pm$20% | | | | |
| $d_{33}$ | 16.9 | 17.1 | 16.2 | 15.8 | 18.1 |
| $d_{32}$ | 4.4 | 4.1 | 4.2 | 3.8 | 3.4 |
| $d_{31}$ | 2.5 | 3.3 | 2.8 | 2.3 | 2.1 |
| | Sellmeier equation:[(a)] $n_i(\lambda)^2 = A_i + B_i/[1-(C_i/\lambda)^2] - D_i\lambda^2;\ i=\{x,y,z\}$ | | | | |
| $A_x$ | 2.11460 | 2.15559 | 2.11055 | 2.22681 | 2.34498 |
| $B_x$ | 0.89188 | 0.93307 | 1.03177 | 0.99616 | 1.04863 |
| $C_x[\mu m]$ | 0.20861 | 0.20994 | 0.21088 | 0.21423 | 0.22044 |
| $D_x[\mu m]^{-2}$ | 0.01320 | 0.01452 | 0.01064 | 0.01369 | 0.01483 |
| $A_y$ | 2.15180 | 2.38494 | 2.38888 | 1.97756 | 2.74440 |
| $B_y$ | 0.87862 | 0.73603 | 0.77900 | 1.25726 | 0.70733 |
| $C_y[\mu m]$ | 0.21801 | 0.23891 | 0.23784 | 0.20448 | 0.26033 |
| $D_y[\mu m]^{-2}$ | 0.01327 | 0.01583 | 0.01501 | 0.00865 | 0.01526 |
| $A_z$ | 2.31360 | 2.27723 | 2.34723 | 2.28779 | 2.53666 |
| $B_z$ | 1.00012 | 1.11030 | 1.10111 | 1.20629 | 1.10600 |
| $C_z[\mu m]$ | 0.23831 | 0.23454 | 0.24016 | 0.23484 | 0.24988 |
| $D_z[\mu m]^{-2}$ | 0.01679 | 0.01995 | 0.01739 | 0.01583 | 0.01711 |

(a) Except for KTP, Sellmeier equations for the isomorphs were obtained from refractive indices data between 0.4–1.5 $\mu$m.

In the former case, a lower melting potassium polyphosphate mixtures (e.g. $K_4P_2O_7$-$2KPO_3$ – commonly referred to as "K6" flux) is used as a high temperature ($\sim$ 950°C) solvent for KTP and the crystal growth process is controlled by slowly cooling the saturated solution over a range of $\sim$ 50°C. Compare to melt growth techniques such as the Bridgeman growth of $AgGaSe_2$ or the crystal pulling of $LiNbO_3$, flux growth is much slower, with a typical crystal growth rate of $\sim$1–1.5mm/side-day. Despite this, large transparent crystals of up to 32 $\times$ 42 $\times$ 87mm$^3$ have been successfully grown. Current crystal growth emphasis is on scaling to larger solution volume and faster crystal growth rate. Recent advances in sophisticate process control equipment have led to significant advances in high temperature flux growth technology. Future progress in the crystal growth of KTP will continue to run parallel to these advances.

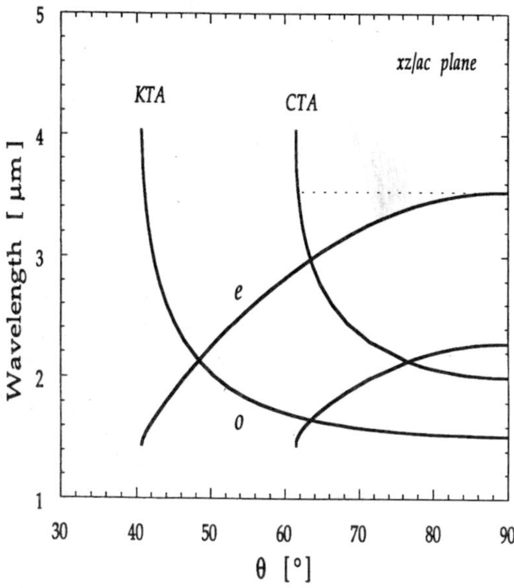


**Fig. 5.9** Calculated Type II, 1.064 μm pumped OPO tuning curves for $KTiOAsO_4$ and $CsTiOAsO_4$. The tuning curves for $RbTiOAsO_4$ lie between these two cases and are not shown for clarity.

The hydrothermal crystallization of KTP is similar to the flux process except water is added to lower the viscosity of the solution and the crystal growth temperature (~550°C). To prevent the water from evaporating, the solution is placed inside a closed high pressure (~25,000psi) vessel. Although the typical hydrothermal crystal growth rate (~1mm/side-week) is slower than flux growth, crystal properties, such as optical damages and insertion loss, of hydrothermally grown crystals appear to be better. Furthermore, recent modification of the solution chemistry has led to a hydrothermal process with significantly lower operating temperature and pressure and higher crystal growth rate.[102,103] The lower operating temperature and pressure make the problem of scaling to larger solution volume technically viable, and is particularly promising for the production of the long crystals needed for OPO use.

### 5.4.2 Potassium Niobate ($KNbO_3$)

Since 1970, potassium niobate has been recognized as a promising efficient crystal with very large second order nonlinearity. Serious crystal growth problems, however, had limited its use in frequency conversion devices. During the past decade, many of these problems were solved and high quality single crystals are now commercially available for device fabrication. The material development of $KNbO_3$ has been recently reviewed by Xing.[104]

Like KTP, $KNbO_3$ crystals are grown by a high temperature flux technique, with the solvent being a small amount of excess $K_2O$. Historically the material development of $KNbO_3$ was plagued by problems of crystal coloration and crystal cracking. During the early crystal development of $KNbO_3$, a blue coloration was often observed which destroyed the optical transparency.[105] The exact nature of this coloration is still unknown, although a recent study[106] concludes that the coloration was due to light scattering rather than impurity related optical absorption. Nevertheless, by confining the starting crystal growth composition to between 52–53 mole % $K_2O$ concentration and improving melt homogenization, transparent water-like crystals of $KNbO_3$ can now be commercially produced.

$KNbO_3$ is cubic at the crystal growth temperature $\sim$1070°C. Upon cooling, it undergoes two separate phase transitions: at $\sim$430°C to tetragonal and at $\sim$200°C to orthorhombic. These transitions often lead to cracking and polydomain formation, which had limited the usable size of early crystals. Metrat and Deguin has shown that crystal cracking can be minimized by careful control of the furnace's temperature gradient.[107] Appropriate post-growth electrical and mechanical polings can then be used to produce single ferroelectric domain crystals suitable for device applications. To date, $KNbO_3$ crystals boules weighing over 100 g have been grown, and fabricated single domain crystals with linear dimensions of 2 cm have been obtained. These advances not withstanding, the phase transitions of $KNbO_3$ are still troublesome, as they make the exact orientation of the seed crystal, and thus, the grown crystal boule, somewhat unpredictable.

The linear and nonlinear optical properties of $KNbO_3$ have been reported in details by Gunter and coworkers in two recent papers[108,109] and our discussion below will be based on this work. Room temperature $KNbO_3$ is orthorhombic (point group *mm2*; space group Bmm2), with $\mathbf{a} = 5.6896\,\text{Å}$, $\mathbf{b} = 3.9692\,\text{Å}$ and $\mathbf{c} = 5.7256\,\text{Å}$ (polar axis). It has the same susceptibility tensor elements ($d_{15}$, $d_{24}$, $d_{31}$, $d_{32}$ and $d_{33}$) as KTP. Also, like KTP, $d_{31}$ and $d_{32}$ are comparable, and have the same sign (see Table 5.4).[110] It is negatively biaxial ($n_\mathbf{b} > n_\mathbf{a} > n_\mathbf{c}$), with a birefringence $n_\mathbf{b} - n_\mathbf{a} \sim 0.42(n_\mathbf{b} - n_\mathbf{c})$ at 0.633 $\mu$m. Since interchanging the **a** and **b** axes corresponds to exchanging ($d_{24}$, $d_{32}$) with ($d_{15}$, $d_{31}$), the ordering of $KNbO_3$'s principal refractive indices is essentially the reverse of KTP's $n_\mathbf{c} > n_\mathbf{b} > n_\mathbf{a}$. Together with $d_{31} \approx d_{32}$, this reverse ordering qualitatively explains the complementary nonlinear optical properties of these two materials. For instance, while KTP allows efficient Type II phase-matched process in its principal planes only, $KNbO_3$ is only effective when used with Type I phase-matching. It is worth noting that, with Kleinman symmetry, the complementary nature of the $d_{eff}$(I) and $d_{eff}$(II) between positive and negative uniaxial crystals is also evident in Table 5.1.

In Table 5.2, the optical dielectric axes $\{xyz\}$ of $KNbO_3$ correspond to the crystallographic axes $\{\mathbf{a}, \mathbf{c}, \mathbf{b}\}$. Thus the $yz$ plane has the largest birefringence, and the corresponding $d_{eff}$ for SHG is given by:

$$d_{eff}(\mathrm{I}) = d_{31}\cos^2\theta + d_{32}\sin^2\theta \qquad (5.16)$$

The reported Type I coupling constants $d_{eff}$ in this plane lies between 15.8–18.3 pm/V. This is 4–9 times larger than $d_{eff}$(II) in KTP, and is ~3 times that of $LiNbO_3$'s $d_{eff}$(I). Although $KNbO_3$ has a larger birefringence than KTP, it is also more dispersive, which limits its Type I SHG cutoffs in the $yz$ plane to ~0.85 and ~0.98 $\mu$m (compared to the 0.799 $\mu$m Type I SHG cutoff of KTP). These properties make $KNbO_3$ a leading candidate for the bulk frequency doubling of diode lasers for optical storage and imaging applications.[111]

Despite the high nonlinearity available in the $yz$ plane, the small birefringence variation makes this plane not particularly useful for parametric application since suitable pump lasers are not available. At room temperature, the longest pump wavelength for parametric down conversion in the $yz$ plane is ~491 nm. Kato has reported efficient oscillation of a 0.532 $\mu$m pumped $z$-cut $KNbO_3$ OPO.[112] Phase matching is achieved by temperature tuning and energy conversion efficiency as high as 32% were reported. Unfortunately, due to the high phase matching temperature (~180°C) needed to reach output wavelength degeneracy, the tuning range of this device is small – between 0.88-1.35 $\mu$m only. Although broader tuning should be possible with a shorter pump wavelength (e.g. the second harmonic of the Ti-sapphire), the 0.6–1.8 $\mu$m outputs overlap with other more readily available laser sources, and are thus not of great interest (Figure 5.10). In addition, it has been reported that temperature cycling can lead to the ferroelectric depoling of $KNbO_3$.[113] Although the re-poling of suitably cut crystals has been demonstrated, crystal depoling does raise serious reliability issue which must be addressed before this material can gain wide acceptance.

Broad tunability, with the idler wave extending to the infrared absorption cutoff, can in principle be obtained by angle tuning in the $xz$ or $xy$ planes with relatively high nonlinear coupling. Figure 5.11 shows the calculated tuning curve for a Type I $KNbO_3$ OPO pumped at 0.532, 0.750 and 1.064 $\mu$m. In these phase-matching schemes, unless a large pump beam diameter is used, beam walkoff can be relatively severe and must be compensated with multiple crystals.[114]

Using expressions in Table 5.2, we estimate that the nonlinear coupling constants are between 11.5 to 15 pm/V for these processes. Similarly, the temperature bandwidth, estimated from published thermo-optic coefficients,[109] is ~0.3°C-cm, suggesting that temperature control would be needed for stable operation. As discussed earlier, in ultrashort pulse generation, the high nonlinearity of $KNbO_3$ allows the use of thin crystals which can render the temperature and angular acceptances unimportant.

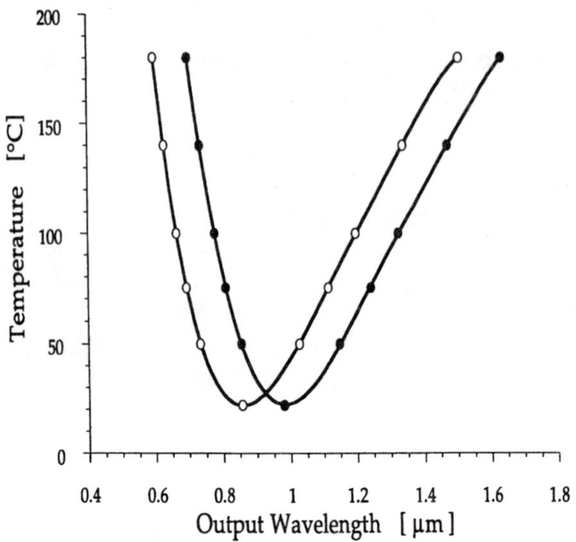


**Fig. 5.10** Temperature tuned $KNbO_3$ OPO tuning curves.[108] (•) is for a 0.491 $\mu$m pump propagating along $z$ (i.e. **b**) axis. (○) is for a 0.428 $\mu$m pump propagating along the $y$-axis (i.e. **a**). Upper temperature is limited by the *orthorhombic* → *tetragonal* phase transition of $KNbO_3$.

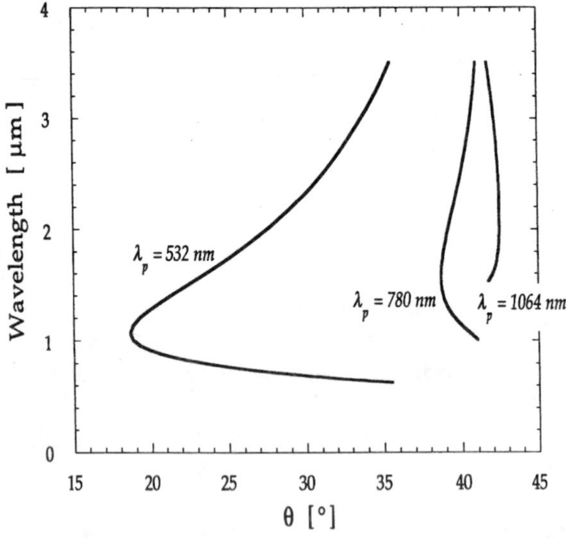


**Fig. 5.11** Type I, angle-tuned OPO outputs of $KNbO_3$. Pump wavelengths ($\lambda_p$) are at 0.532, 0.780, and 1.064 $\mu$m, and are propagating in the $xz$/**bc** plane.

From Figure 5.11, we see that a Ti-sapphire (780 nm) pumped $KNbO_3$ OPO offers an extremely broad spectral bandwidth near its wavelength degeneracy $\sim$ 1.56 $\mu$m. The calculated reciprocal group velocity mismatch between the pump and signal is a modest ~50 fsec/mm for $KNbO_3$, and its nonlinear gain ($d_{\text{eff}}^2/n^3$) is 10 times higher than KTP. Since both the signal and idler outputs have the same polarization at degeneracy, this degenerate $KNbO_3$ OPO may be particularly suitable for the cascade pumping of another OPO stage (e.g. $AgGaS_2$). Overall, although $KNbO_3$ has not been widely studied as an OPO crystal, its known optical properties suggest that it should be a promising material for parametric generation in the near infrared.

## 5.5 VISIBLE AND UV OPO MATERIALS

### 5.5.1 $\beta$-Barium Metaborate (BBO)

The composition barium metaborate ($BaB_2O_4$) has been known since 1900. It is widely used as a low melting flux for crystal growth and has been used as a fungicide additive in paints. Early investigations of the crystal structure of the high and low temperature phases were reported by Mighell[65] and Hubner[115] respectively prior to 1970. The usefulness of the low temperature phase ($\beta$) crystal for nonlinear optics, however, was not recognized until 1980, when it was proposed that the accentric, planar $(B_3O_6)^{-3}$ ring structural unit should be an ideal candidate for 'engineering' nonlinear optical materials useful for generating ultraviolet radiation.[116] Recognizing the mechanical and chemical fragility of organic nonlinear crystals, and the lack of robust near-UV frequency doubler, Chen and coworkers[116] initiated a systematic investigation of the nonlinear optical properties of borate crystals in early 1980. This effort eventually led to the discovery of BBO and the LBO crystals for nonlinear optics.

$\beta$-barium metaborate (BBO) belongs to the point group $3m$, (space group $R3c$). The crystal structure is made up of layers of planar $(B_3O_6)^{-3}$ rings stacked along the polar **c**-axis. Figure 5.12 illustrate this stacking[34] and defines the standard crystallographic axes for the crystal. Treating this relatively simple structure as an inorganic example of an oriented 'molecular' gas, Chen[117] and others[118] have theoretically correlated the nonlinearity of BBO to the $\pi - \pi^*$ transition of the conjugated $(B_3O_6)^{-3}$ rings. The strongly anisotropic structure of BBO is also thought to be responsible for the large crystal birefringence ($n_e - n_o \sim 0.11$) that makes it so useful in the ultraviolet. As shown in our example earlier (Equation (5.4)), the effective nonlinear coupling in BBO varies as $\sim \cos\theta$, and becomes vanishingly small near 90° phase matching. As crystal birefringence reaches a maximum at 90°, the $\sim \cos\theta$ dependence is thus 'incompatible' with phase-matching requirements near the UV cutoff. Fortunately, BBO's large crystal birefringence compensates for this weakness, and allows for

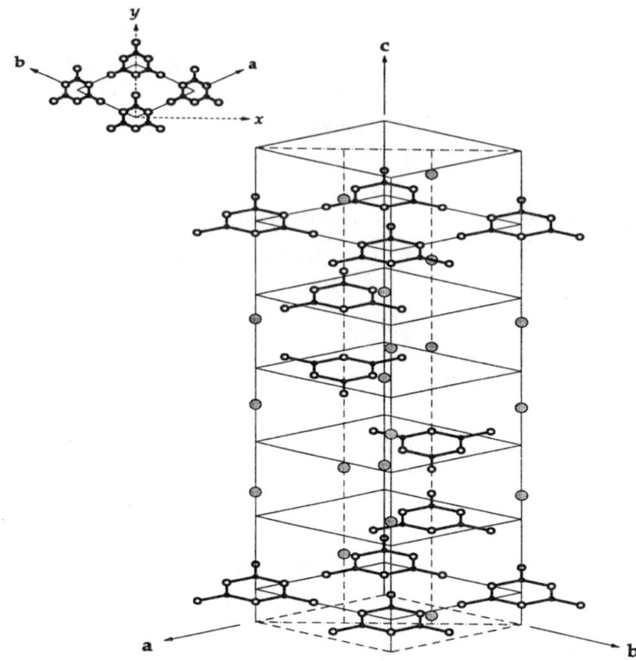


**Fig. 5.12** Crystal structure of $\beta$-barium borate showing the stacking of planar graphite-like structure. This planar stacking is thought to be responsible for the weak $\{00\cdot1\}$ basal cleavage.

efficient phase-matching of most OPO processes within a relatively small angle range (i.e. $25^\circ < \theta_{pm} < 50^\circ$).

BBO is the first nonlinear crystal which met all the stringent material requirements for efficient OPO operation in the visible and near ultraviolet. Compare to urea which BBO replaces, BBO is superior in all respects (see Table 5.6).[16,17,34,119,120] It is mechanically robust, chemically stable and has a broader transparency (0.2–3.3 $\mu$m) than urea both in its UV and infrared cutoffs. In addition, its crystal growth is much easier than urea's and it does not suffer from multishot optical damage problem. Given these outstanding properties, it is easy to understand how BBO has replaced urea crystal for most OPO applications.

In Section 5.1, we have derived the $d_{eff}$ for Type I interaction in BBO (see Equations (5.1a), (5.1b) and (5.4)). From Table 5.1, we see that BBO is also efficient for Type II SHG phase matching:

$$d_{eff}(\mathrm{I}) \approx -d_{22} \cos\theta \sin 3\phi \quad \text{for } o + o \rightarrow e \tag{5.17}$$

$$d_{eff}(\mathrm{II}) \approx d_{22} \cos^2\theta \cos 3\phi \quad \text{for } o + e \rightarrow e \tag{5.18}$$

**Table 5.6** Properties of selected nonlinear crystals for near-UV and visible OPO's.

| Crystals | Urea | $\beta$-$BaB_2O_4$ | $LiB_3O_5$ [(a)] |
|---|---|---|---|
| point group | $\bar{4}2m$ | 3m | *mm2* |
| cell constants | **a** = 5.645 Å<br>**c** = 4.704 Å | **a** = 12.548 Å<br>**c** = 12.737 Å | **a** = 8.4473 Å<br>**b** = 7.3788 Å<br>**c** = 5.1395 Å |
| transparency [$\mu$m] | 0.2–1.4 | 0.19–3.3 | 0.16–3.3 |
| efficient phase matching configuration | $o \to e+e$<br>$o \to o+e$ | $e \to o+o$<br>$e \to e+o$ | $e \to o+o(xy)$<br>$e \to o+e(xz)$<br>$o \to o+e(yz)$ |
| tuning | angle | angle | temp./angle |
| refractive indices @ 0.532 $\mu$m | $n_o$=1.4856<br>$n_e$=1.6100 | $n_o$=1.6749<br>$n_e$=1.5555 | $n_{\mathbf{b}}$=1.6212<br>$n_{\mathbf{c}}$=1.6064<br>$n_{\mathbf{a}}$=1.5787 |
| nonlinearity[(b)] $d_{ij}$ [pm/V] | $d_{14}$=1.4 | $d_{22}$=2.2<br>$d_{31}$~0.1 | $d_{15}$~$d_{31}$ [(c)]<br>$d_{24}$~$d_{32}$<br>$d_{31}$=1.05<br>$d_{32}$=0.98<br>$d_{33}$=0.05 |
| $\Delta T.\ell$[°C-cm] | ~30 | ~50 | ~3–15 |
| crystal damage threshold @ 355 nm [GW/cm$^2$] | 0.18 | 10 | 19 |
| 355 nm pumped OPO | | | |
| phase matching type | II | I | I $(xy)$ |
| "tuning range" [$\mu$m] | 0.498–1.23 | 0.4–2.7 | 0.4–2.7 |
| $\theta_{pm}$ range [°] | 50–90 | 22–33 | $\phi$~ 15–45 |
| walkoff $\rho$[°] | -(0–5) | ~4 | ~1 |
| 266 nm pumped OPO | | | |
| phase matching type | II | I | II |
| tunable UV [$\mu$m] | 0.33–0.4 | 0.29–0.4 | 0.29–0.35 |
| $\theta_{pm}$ range [°] | 45–70 | 27–40 | 0–27 $(xz)$<br>$\varepsilon$ 0–90 $(yz)$ |
| crystal growth | solution | flux | flux |
| crystal sizes [cm] | ~3–5 | 2 | 2 |

(a) for $LiB_3O_5$, the crystallographic {**a**,**c**,**b**} axes correspond to dielectric $\{x,y,z\}$ axes.
(b) $ij$ refer to piezoelectric {X, Y, Z} axes, except for LBO where it denotes the crystallographic {**a**,**b**,**c**} axes. $d_{ij}[m/V]=\frac{4\pi}{3}\times10^{-4}d_{ij}[esu]$.
(c) by Kleinman symmetry.

The $\cos\theta$ dependence suggests that only critical phase matching can be efficient in BBO. The design of BBO OPO's will thus have to address the walkoff problem if very high conversion efficiencies are to be achieved.[114]

Optical parametric oscillators using both Type I and Type II interaction in BBO have been reported.[114,121,122] The majority of these devices have been pumped by the third harmonic of the Nd:YAG laser to give tunable visible and near IR outputs, although 308 nm excimer laser pump has also been used. While Type I OPO's offer higher parametric gain than Type II, it has significantly broader spectral bandwidths (see Figure 5.11), making it less suitable for spectroscopic applications. The $\cos^2\theta$ dependence in Eq. (5.18) means a faster angular drop off of $d_{\text{eff}}(\text{II})$ than $d_{\text{eff}}(\text{I})$. In addition, as Type II phase matching requires about twice as much birefringence as Type I to phase match the same three waves, a larger phase matching angle $\theta$ is needed, further lowering the efficiency. Fortunately this lower nonlinear gain can be partially compensated by resonating the extraordinary wave. As both the pump and the signal are extraordinary, birefringence walk off problems are reduced, and longer crystals may be used.

Pumped at 266 nm, BBO OPO can provide broadly tunable radiation from 0.3–2.6 $\mu$m. The practical utility of this OPO is, however, limited by two factors. First, the multistack dielectric coatings currently used for the trichroic OPO mirrors have low ultraviolet damage resistance. Second, the quadrupling of Nd:YAG output is often inefficient, and the poor UV beam quality further aggravates optical damage problems and limits efficient conversion. Using intracavity pump steering mirrors, Bosenberg and coworkers[123] have bypassed the first limitation and demonstrated efficient parametric oscillation of a 266 nm pumped BBO OPO. With respect to the second limitation, the recently developed high power, intracavity frequency doubled Nd-lasers ("Green-YAG") should allow the efficient generation of high quality 266 nm beams suitable for pumping OPOs.

Commercial production of BBO crystals currently uses the high temperature solution growth technique, employing either $Na_2O$ or $Na_2B_2O_4$ fluxes.[22,124,125] The top-seeded pulling approach, which is analogous to the Czochralski crystal pulling of silicon, is used in these systems in order to suppress the anomalously fast growth of BBO along the **c**-direction.[126] As grown crystals are short [00·1] disc-like boules as shown in Figure 5.13. Since phase matched parametric interaction occurs at $25^\circ < \theta_{\text{pm}} < 50^\circ$, the disc-like geometry of these crystal boules limits the available crystal interaction length. Top-seeded pulling along the ~ [02·1] direction places the phase-matching direction along the diameter of the disc-like boule and can in principle produce significantly longer crystals. Unfortunately, the graphite-like, planar stacking crystal structure leads to a relatively weak basal cleavage {00 · 1}, which can easily crack when subjected to the high temperature gradient of the crystal growth furnace.[22,127] Nevertheless, by fine tuning the radial temperature gradient, it is possible to grow inclusion-free ~ [02·1] boules with minimal cracking.[128]

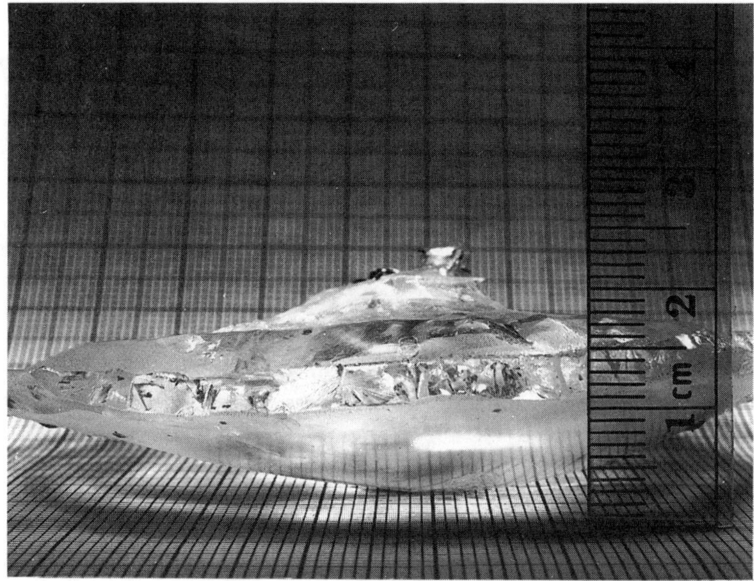


**Fig. 5.13** Top-seeded solution-grown crystal boules of $\beta$-$BaB_2O_4$ (grown at Cornell University).

This has led to significantly longer (~2 cm) crystals especially suitable for Type II parametric oscillation.[114]

In order to obtain high quality crystals, the crystal growth rate during commercial production has been limited to ~0.2–0.5 mm/day. This is due to the high viscosity of the borate flux, and its poor crystal growth kinetics. The primary defect in BBO is microscopic flux inclusions which lead to scattering loss and lower the damage thresholds.[34] Although other less viscous fluxes, such as $BaCl_2$ and $BaF_2$, have been suggested, careful comparison of the practical usefulness of these fluxes has not been reported.

Two particularly promising recent crystal growth advances are worth mentioning. First by taking advantage of the supercooling tendency of barium borates melt, Kouta and coworkers[129] have demonstrated that large (15 mm diameter × 40 mm) $\beta$-phase crystals can be grown by direct Czochralski pulling from a stoichiometric melt. Besides a 10-fold increase in crystal growth rate, this technique also yields crystal boules with an aspect ratio especially suitable for fabricating OPO crystals. Whereas Kouta *et. al.* pushes the upper limits of the temperature and thermal gradient, Bordui and coworkers have taken just the opposite approach and have recently demonstrated that large, blocky and inclusion-free BBO crystals (>5 cm along **c**) can be grown under isothermal condition by slow cooling a low melting (~800°C) BBO-NaCl solution.[130] While the nonlinear optical performance of these large crystals have not yet been reported, these new growth techniques will likely be instrumental in bringing about the commercialization of visible and near UV OPO technology.

### 5.5.2 Lithium Triborate (LBO)

The $(B_3O_6)^{-3}$ ring in BBO can be viewed as a planar linkage of three smaller $(BO_3)^{-3}$ structural units with three shared-oxygen atoms. Another structural unit commonly found in boron-oxygen compounds is the tetrahedral $(BO_4)^{-5}$ group. In a boron-oxygen system, $(BO_3)^{-3}$ and $(BO_4)^{-5}$ units can readily link with each others to form extended polymeric networks. This polymeric linkage not only accounts for the glass-like properties of molten borate melt, it is also responsible for the structural diversity which exists among inorganic borate compounds. In 1989, the systematic search for optical nonlinearity among the borates at the Fujian Institute yielded a new group of promising boron-oxygen compounds, namely $LiB_3O_5$, $CsB_3O_5$ and $TlB_3O_5$.[131] Like BBO, these triborates are made up of hexagonal boron-oxygen rings, in this case $(B_3O_7)^{-5}$, which are responsible for the observed optical nonlinearity. The $(B_3O_7)^{-5}$ ring is formed by linking two $(BO_3)^{-3}$ units with one $(BO_4)^{-5}$ unit. By adding an extra oxygen to the ring, the $\pi$-conjugation is destroyed and a non-planar structural building block results. In the triborates, the $(BO_4)^{-5}$ group serves as bridging units among neighboring $(B_3O_7)^{-5}$ rings, resulting in extended helical chains which make up the crystal structures.[132] The strong anisotropy of the planar stacking BBO structure is now absent, which qualitatively explains the substantially reduced birefringence observed in the triborates (e.g. $LiB_3O_5$).

Of the three triborates, only the lithium compound has been widely studied. LBO is orthorhombic with *mm2* point symmetry (space group $Pna2_1$, with $\mathbf{a} = 8.4473$ Å, $\mathbf{b} = 7.3788$ Å and $\mathbf{c} = 5.1395$ Å (polar axis)). It is negatively biaxial ($2\Omega \sim 109°$ @ 540 nm), with $n_\mathbf{b} > n_\mathbf{c} > n_\mathbf{a}$. The crystal birefringence $n_\mathbf{b} - n_\mathbf{a}$ ~0.04 (@ 1.064 $\mu$m) is small, only ~40% that of BBO. Thanks to its unusually good UV transparency (cutoff @ ~0.16 $\mu$m *vs.* 0.165 $\mu$m of fused quartz and 0.19 $\mu$m of BBO), LBO has a small dispersion, which compensates for the small birefringence available for phase matching. This low dispersion also provides broad angular, temperature and spectral bandwidths for most phase-matching processes in LBO. For SHG in the visible and near UV its broad angular bandwidths makes it particularly attractive for frequency mixing of $Nd^{3+}$ lasers and its harmonics[133] and for frequency doubling the Ti-sapphire laser.[134]

The infrared absorption cutoff of LBO is at 3.3 $\mu$m, which is preceded by broad overtones beginning at ~2.6 $\mu$m. This makes it more suitable as a UV-visible material and we have classified it as such. Nevertheless, because it has *mm2* point group symmetry, it is useful to compare LBO with the two orthorhombic visible-near IR crystals which we have discussed earlier. In a way, the nonlinear optical properties of LBO can be regarded as a compromise between the two extremes represented by KTP and $KNbO_3$. Like KTP and $KNbO_3$, although $d_{31}$ and $d_{32}$ of LBO are not related by any known symmetry, they are comparable in magnitude (see Table 5.5). Yet unlike these two *mm2* crystals, $d_{31}$ and $d_{32}$ have opposite signs. With the polar **c**-axis lying along the optical $y$-axis rather than in the $xz$ plane, LBO is efficient for both Type I and II phase matching (see Table 5.2). This of course adds to LBO's versatility in device applications. Taking advantage of this property, researchers have demonstrated efficient operation of both Type I and Type II angle tuned OPO's pumped at the 3rd and 4th harmonics of the Nd:YAG,[135,136] and at the 308 nm XeCl excimer line.[137] Figure 5.14 shows the tuning curves of these devices calculated from the Sellmeier equations given in Reference 133.

Unlike $KNbO_3$ and KTP, the birefringence in the $yz$ plane of LBO is ~35% of the total birefringence. Thus, LBO lacks the advantageous quasi-uniaxial property which KTP enjoys. On the other hand, LBO's low dispersion and modest thermo-optical coefficients permits temperature tuned non-critical phase matching and temperature tuning is achieved without significant degradation in its temperature bandwidth. Compared to KTP's ~(20–175)°C-cm temperature bandwidth, and $KNbO_3$'s $\Delta T.\ell \sim 0.3$°C-cm (see Table 5.4 and 5.6), LBO's ~(3–15)°C bandwidth represents a realistic compromise. With NCPM temperature tuning,[131,138] LBO can easily prove to be more efficient than BBO in applications where walk-off and beam divergence problem must be minimized.

The availability of high quality crystals also makes possible the direct experimental comparison of OPO performance using nonlinear crystals with different phase-matching properties.[100,139,140] From these experiments, it was concluded that in critically phase-matched and angle tuned OPO's, pump-signal walk-off and off-axis reflection loss often lead to significantly higher oscillation threshold than in NCPM and temperature tuned devices. In this regard, LBO offers significant advantage over BBO. Given its modest temperature acceptance, LBO's temperature tuning range is surprisingly broad, with infrared output at 2.6 $\mu$m readily attainable at a comparatively low temperature of 190°C. Ebrahimzadeh *et. al.* have demonstrated the efficient operation (~20% pump depletion) of an all solid-state, Q-switch mode-locked LBO OPO tunable between 0.65–2.65 $\mu$m.[141] Significantly improved efficiency and spectral bandwidth of this device and its extension into short wavelengths have also been reported.[142,143,144]

Inclusion-free LBO crystals have been found to have a high optical damage threshold, which has been reported to be twice that of BBO.[16] Although the reason for this high damage threshold still remains unclear, it was suggested that the absence of transition metal impurities in LBO are primarily responsible.[131] It is believed that the small boron and lithium atoms lead to a 'compact' crystal structure that effectively

1.5
1.0
0.5
0.0
$-d_{eff}$ [pm/V]
0
30
60
90
3
2
1
0
Wavelength [μm]
$\lambda_p$=532 nm
$\lambda_p$=355 nm
$\lambda_p$=266 nm
xy/ac plane
0
30
60
90
φ [°]

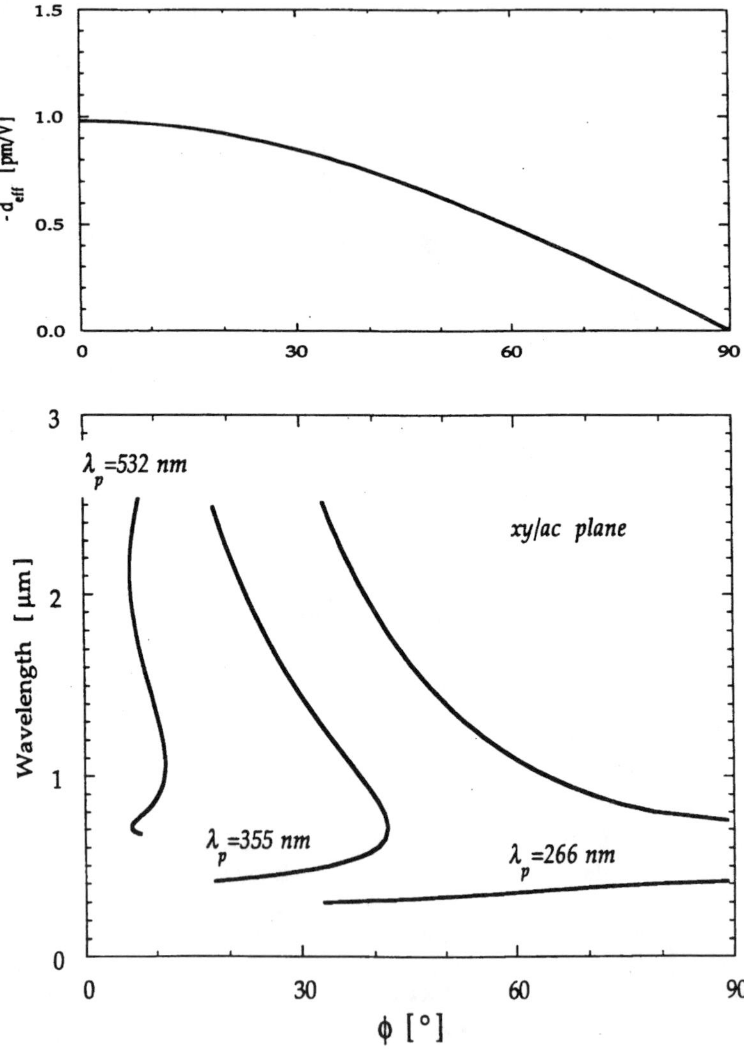

a

Fig. 5.14

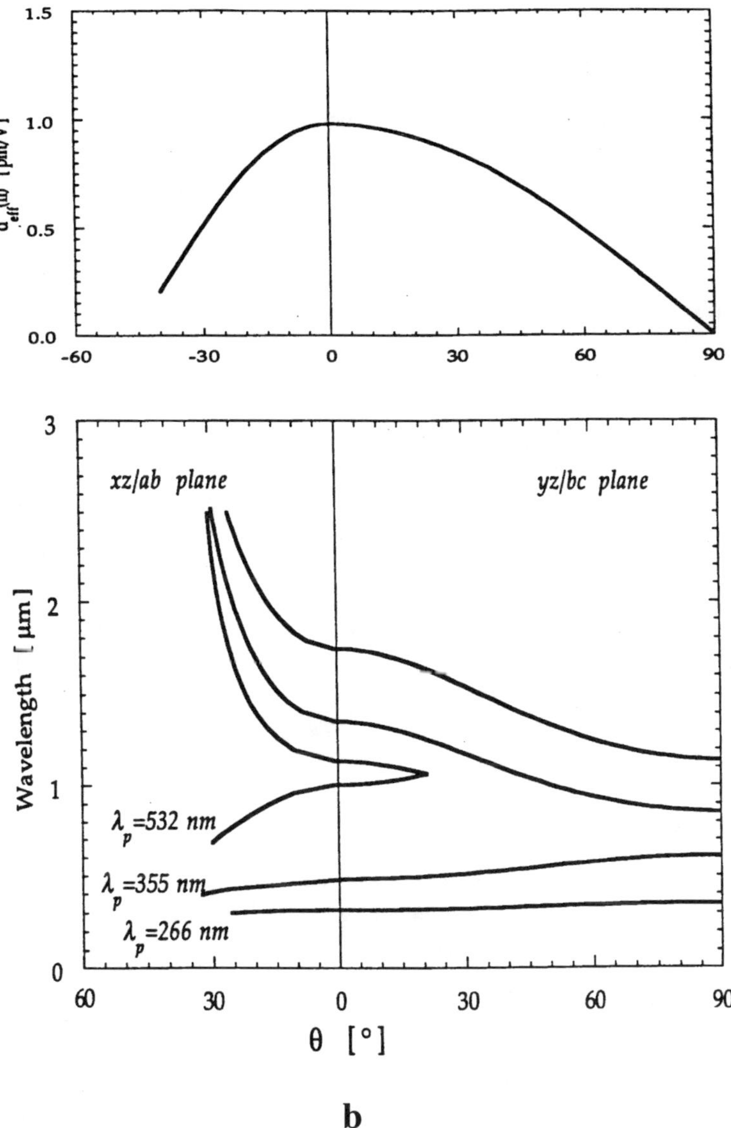


**b**

**Fig. 5.14** $d_{\text{eff}}$ for SHG and tuning characteristics of $LiB_3O_5$ OPO's pumped by the second, third and fourth harmonics of Nd:YAG. (a) Type I phase-matching in $xy$/**ac** plane, and (b) Type II, in $xz$ and $yz$ planes. For clarity, the 0.532 $\mu$m pumped tuning curve in (b) is arbitrarily truncated at its wavelength degeneracy.

keeps out larger transition metal ions which can act as damage centers.[145] The short UV absorption edge of LBO may also be responsible for the high damage threshold. In an OPO pumped by the third harmonic of Nd:YAG, LBO's 160 nm absorption cutoff is actually ~18 nm shorter than the two-photon energy of the 355 nm pump. As the damage fluence of LBO is comparable to that of the fused silica optical elements that make up the OPO/OPA system, laser damage to the nonlinear crystal is no longer the primary concern. Although direct measurements have not been made, its thermal properties, including heat induced crystal cracking and thermal dephasing, appear to be excellent.[146,147] These properties have made LBO the premier crystal for very high peak and average power frequency mixing applications. They also make LBO ideally suitable for the parametric generation of intense, tunable picosecond pulses using the single pass OPG/OPA configurations. With a 355 nm pump intensity of ~5 GW/cm$^2$, a remarkable pump-to-signal energy conversion efficiency of 28% has been observed from two 15 mm long LBO crystals.[148,149] Compared to OPO's, single pass OPG/OPA devices not only offer design simplicity and efficient useful energy output coupling, they completely bypass limitation imposed by the low damage thresholds of dielectric mirror coatings. Extension of these OPG/OPA devices to shorter (e.g. 266 nm) pump wavelengths should prove particularly interesting.

Except for its relatively small nonlinearity, LBO possesses near ideal characteristics for OPO applications. Its broad transparency, high damage thresholds, broad temperature, angular and spectral bandwidths, temperature tunability and efficient Type I and II phase matching all contribute to the crystal's versatility in parametric frequency conversion applications. Unfortunate, the crystal growth technology for LBO currently lags behind BBO. Like BBO, LBO is commercially grown by the top-seeded solution growth technique, with the flux being excess $B_2O_3$. LBO crystal boules measuring ~$35 \times 30 \times 15$ mm$^3$ have been reported.[131] Broad-based applications of OPO's, however, require that high quality, fabricated crystals of 2–3 cm long be readily available. It has recently been reported that LiF additive effectively lowers the viscosity of the lithium borate melt and improves the crystal growth process.[132] This should prove promising for the development LBO-based OPO devices.

## 5.6 ADDITIONAL REMARKS

Since the recent development in OPO technology was primarily driven by the development of new efficient nonlinear crystals, we have focused our discussion only on a few of these crystals. As a result, many less well developed attractive materials have been neglected. Cases in point are the $d$-LAP and analogs[150] and GaSe,[64] which remain promising materials for OPO applications. By focusing on only a few crystals, we seek to illustrate that besides the optical nonlinearity, other crystal properties are equally important in determining the practical usefulness of parametric devices. The need to meet a broad range of important property requirements, including low optical dispersion, high damage threshold, favorable crystal growth properties...etc, explains why truly useful nonlinear crystals are so difficult to develop.

While we have only discussed material issues relating to the nonlinear crystal in this chapter, we must point out that advances in solid-state laser crystal development also play a major role in advancing OPO technology. For instance, recent rapid progress in tunable femtosecond OPO technology was in part triggered by the development of high power femtosecond Ti-sapphire lasers. Similarly, the maturing high power diode lasers technology was responsible for the demonstration of all solid-state, tunable OPO sources. To date, many promising new lasers such as the alexandrite, LiSAF and LiCAF and Yb:YAG, have not been widely used as OPO pump sources. The incorporation of these new materials into commercial laser systems should offer new opportunities for improving the operating characteristics and extending the tuning range of existing parametric devices.

Advances in laser technology are also not necessarily limited to new or better laser crystals. For instance, the commercial development of injection seeded, single longitudinal mode Nd:YAG lasers are clearly instrumental to the recent development of pulsed nanosecond OPO's. Similarly, the passive negative feedback mode-locked (PNFM) Nd:YAG laser recently developed by Del Corno and coworkers[151] has a temporal pulse structure ideally suitable for pumping OPO's. The usefulness of the PNFM laser as an OPO pump source was subsequently verified in a *non-collinearly* pumped LBO OPO, which yielded single pulse conversion efficiency as high as 70%.[152] Lastly, the development of feedback stabilization techniques useful for frequency standard applications will undoubtedly benefit the development of stable, broadly tunable cw parametric oscillators.[94]

These comments notwithstanding, we still believe that optical parametric devices will remain a materials-driven technology. As OPO/OPA devices mature from laboratory curiosity to real-world applications, more emphasis will be placed on higher useable output energy, better beam quality (both spatial and spectral), device compactness, reliability, ease of operation and low cost. To meet these requirements, new and longer crystals with higher parametric gain and improved phase-matching characteristics will be needed. Given the slow, tedious and unpredictable nature of materials development, it is likely that the development of nonlinear crystals and laser crystals will remain as the bottle-neck for future progress in OPO technology. In the next chapter, we summarize recent results on nanosecond and femtosecont OPO's.

## REFERENCES

1. L.G. Koreneva, V.F. Volin, B.L. Davidov, *Molecular Crystals in Nonlinear Optics*, "Nauka", Moscow, 1975; D.S. Chemla and J. Zyss, *Nonlinear Optical Properties of Organic Molecules and Crystals*, (Academic Press: 1987).

2. S.K. Kurtz and T.T. Perry, *J. of Appl. Phys.*, **39**, 3798 (1968).
3. S. Velsko, *Proc. SPIE*, **681**, (1986) 25.
4. R.S. Feigelson and R.K. Route, *Opt. Eng.*, **26**, 113 (1987).
5. P.A. Budni, P.G. Schunemann, M.G. Knights, T.M. Pollak and E.P. Chicklis, in *Topical Meeting on Adv. Solid-State Lasers*, Hilton Head, South Carolina, **12**, 356 (OSA: 1991).
6. D.C. Edelstein, *New sources and techniques for ultrafast lasers spectroscopy*, Cornell University, *Ph. D. Thesis*, (1990).
7. E.C. Cheung and J.M. Liu, *J. Opt. Soc. Am. B*, **8**, 1491 (1991).
8. J. Chung and A.E. Siegman, *J. Opt. Soc. Am. B*, **10**, 2201 (1993).
9. P.P. Bey and C.L. Tang, J. Quant. Elect. QE-8, 362–369 (March, 1972); W.S. Pelouch, P.E. Powers and C.L. Tang, *Opt. Lett.*, **17**, 1070 (1992).
10. J.D. Kafka, M.L. Watts and J.W. Pieterse, in *Postdeadline paper in the Conference on Lasers and Electro-Optics*, Baltimore, Maryland, CPD32 (OSA: 1993).
11. N. Bloembergen, *IEEE J. of Quantum Electronics*, **QE-10**, 375 (1974).
12. M.G. Roelofs, *J. Appl. Phys.*, **65**, 4976 (1989).
13. R.L. Byer in *Quantum Electronics: A Treatise – Nonlinear Optics*, **B**, H.T. Rabin and C.L. Tang, *Eds.*, 588 (Academic Press: 1975).
14. D. Guyer, C. Hamilton, F. Braun, D. Lowenthal and J. Ewing, in *Proceedings of the Conference on Lasers and Electro-Optics*, Baltimore, MD, (Optical Society of America: 1991) CPDP6.
15. F. Ahmed, *Appl. Optics*, **28**, 119 (1989).
16. S. Lin, Z. Sun, B. Wu and C.T. Chen, *J. Appl. Phys.*, **67**, 634 (1990).
17. L.K. Cheng, M.J. Rosker and C.L. Tang in *Topics in Applied Physics*, **59**, L.F. Mollenauer, J.C. White and C.R. Pollock, *Eds.*, 209 (Springer-Verlag: 1992).
18. S.J. Brosnan and R.L. Byer, *IEEE J. of Quantum Electron.*, **QE-15**, 415 (1979).
19. F. Ahmed, R. Belt and Gashurov, *Appl. Phys. Lett.*, **60**, 839 (1986).
20. K. Kato, *IEEE J. of Quantum Electron.*, **27**, 1137 (1991).
21. D.T. Hon and H. Bruesselbach, *IEEE J. of Quantum Electron.*, **QE-16**, 1356 (1980).
22. L.K. Cheng, W. Bosenberg and C.L. Tang, *J. of Crystal Growth*, **89**, 553 (1988).
23. D.Z. Shen, *Prog. Crystal Growth and Charact.*, **20**, 161 (1990).
24. R.L. Byer, J.F. Young and R.S. Feigelson, *J. of Appl. Phys.*, **41**, 2320 (1970).
25. A. Yokotani, A. Miyamoto, T. Sasaki and S. Nakai, *J. of Crystal Growth*, **110**, 963 (1991).
26. R.C. Eckardt, X.Y. Fan, R.L. Byer, R.K. Route, R.S. Feigelson and J. van der Laan, *Appl. Phys. Lett.*, **47**, 786 (1985).
27. D.R. Suhre, *Appl. Phys.* , **B52**, 367 (1991).
28. D.H. Jundt, M.M. Fejer and R.L. Byer, *IEEE J. Quantum Electronic*, **QE-26**, 135 (1990).
29. Y.F. Zhou, J.C. Wang, P.L. Wang, L.A. Tang, Q.B. Zhu, W.Y.A. and H.R. Tan, *J. of Crystal Growth*, **114**, (1991).
30. S. Kan, M. Sakamoto, Y. Okano, H. Keigo and T. Fukuda, *J. of Crystal Growth*, **119**, 215 (1992).
31. J.E. Midwinter and J. Warner, *Brit. J. Appl. Phys.*, **16**, 1135 (1965).
32. *Proc. IRE*, **37**, 1378 (1949).
33. F. Zernike and J.E. Midwinter, *Applied Nonlinear Optics*, (John Wiley & Sons, Inc.: 1973).
34. D. Eimerl, L. Davis, S. Velsko, E.K. Graham and A. Zalkin, *J. Appl. Phys.*, **62**, 1968 (1987).
35. G.D. Boyd, R.C. Miller, K. Nassau, L. Bond and A. Savage, *Appl. Phys. Lett.*, **5**, 234 (1964).
36. J.Q. Yao and T.S. Fahlen, *J. Appl. Phys.*, **55**, 65 (1984).
37. J. Yao, W. Sheng and W. Shi, *J. Opt. Soc. Am. B*, **9**, 891 (1992).
38. H. Ito, H. Naito and H. Inaba, *J. of Appl. Phys.*, **46**, 3992 (1975). (See also references 39 and 40 for typographic corrections.)
39. H. Ito, *J. Appl. Phys.*, **74**, 5918 (1993).
40. J. Shi and J. Bahaeddin, *J. Appl. Phys.*, **74**, 5916 (1993).

41. V.G. Dmitriev and D.N. Nikogosyan, *Opt. Comm.*, **95**, 173 (1993).
42. H.V. Hobden, *J. of Appl. Phys.*, **38**, 4365 (1967).
43. M. Born and E. Wolf, *Principles of Optics*, (Pergamon Press: 1980).
44. B. Boulanger and G. Marnier, *Opt. Commun.*, **72**, 144 (1989).
45. D. Eimerl, S. Velsko, L. Davis, F. Wang, G. Loiacono and G. Kennedy, *IEEE J. of Quantum Electron.*, **25**, 179 (1989).
46. G. Marnier and B. Boulanger, *Opt. Commun.*, **72**, 139 (1989).
47. W.R. Bosenberg, L.K. Cheng and J.D. Bierlein, in *Topical Meeting on Advanced Solid State Lasers*, New Orleans, (Optical Society of America: 1993) **2**, 134.
48. D.Y. Stepanov, V.D. Shigorin and G.P. Shipulo, *Sov. J. Quantum Electron.*, **14**, 1315 (1984).
49. D.A. Roberts, *IEEE J. of Quantum Electron.*, **QE-28**, 2057 (1992).
50. S.R. Marder, J.E. Sohn and G.D. Stucky, *Materials for Nonlinear Optics – Chemical Perspectives*, Vol. **455**, American Chemical Society Symposium Series (American Chemical Society: Washington, D.C., 1991).
51. D. Jossee, S.X. Dou, J. Zyss, P. Andreazza and A. Perigaud, *Appl. Phys. Lett.*, **61**, 121 (1992).
52. J.D. Bierlein, L.K. Cheng, Y. Wang and W. Tam, *Appl. Phys. Lett.*, **56**, 423 (1990).
53. H. Hyuga, C. Goto, Y. Okazaki, A. Harada, S. Mitsumoto, K. Kamiyama and S. Umegaki, in *Topical Meeting in Compact Blue-Green Lasers*, New Orleans, La., (Optical Society of America: 1993) **2**, 382.
54. R.T. Bailey, G. Gourhill, F.R. Cruickshank, D. Pugh, J.N. Sherwood and G.S. Simpson, *J. Appl. Phys.*, **73**, 1591 (1993).
55. T. Ukachi, T. Shigemoto, H. Komatsu and T. Sugiyama, *J. of Opt. Soc. Am. B*, **10**, 1372 (1993).
56. D. Eimerl, S. Velsko, L. Davis and F. Wang, *Prog. Crystal Growth and Charact.*, **20**, 59 (1990).
57. S.B. Monaco, L. Davis, S.P. Velsko, F.T. Wang, D. Eimerl and A. Zalkin, *J. of Crystal Growth*, **85**, 252 (1987).
58. G. Xing, M. Jiang, Z. Shao and D. Xu, *Chin. Phys.-Lasers*, **14**, 357 (1987).
59. N. Zhang, M.H. Jiang, D.R. Yuan, D. Xu, X.T. Tao and Z.S. Shao, *J. of Crystal Growth*, **102**, 581 (1990).
60. W.R. Donaldson and C.L. Tang, *Appl. Phys. Lett.*, **44**, 25 (1984).
61. B.C. Ziegler and K.L. Schepler, *Appl. Opt.*, **30**, 5077 (1991).
62. G.W. Iseler, *J. Crystal Growth*, **41**, 146 (1977).
63. R.S. Feigelson and R.K. Route, *J. of Crystal Growth*, **104**, 789 (1990).
64. K.L. Vodopyanov, L.A. Kulevskii, V.G. Voevodin, A.I. Gribenyukov, K.R. Allakhverdiev and T.A. Kerimov, *Opt. Commun.*, **83**, 322 (1991).
65. A.D. Mighell, A. Perloff and S. Block, *Acta Crystallogr.*, **20**, 819 (1966).
66. M.D. Ewbank, P.R. Newman, N.L. Mota, S.M. Lee, W.L. Wolf, A.G. DeBell and W.A. Harrison, *J. Appl. Phys.*, **51**, 3848 (1980).
67. N.B. Singh, T. Henningsen, Z.K. Kun, K.C. Yoo, R.H. Hopkins and R. Mazelsky, *Prog. Crystal Growth and Charact.*, **20**, 175 (1990).
68. H. Kildal and J.C. Mikkelsen, *Opt. Commun.*, **9**, 315 (1973).
69. G.C. Bhar, L.K. Samanta, D.K. Ghosh and S. Das, *Sov. J. Quantum Electron.*, **17**, 860 (1987).
70. D.N. Nikogosyan, *Sov. J. Quantum Electron.*, **7**, 1 (1977).
71. P.G. Schunemann, P.A. Budni, M.G. Knights, T.M. Pollak, E.P. Chicklis and C.L. Marquardt, in *Topical Meeting on Advanced Solid State Lasers*, New Orleans, LA., (Optical Society of America: 1993) **2**, 131.
72. K.C. Yoo, R.P. Storrick, T. Henningsen, J.A. Spitznagel and R.H. Hopkins, *J. of Crystal Growth*, **125**, 208 (1992).
73. Y.X. Fan and R.L. Byer, *SPIE – New Lasers for Analytical and Industrial Chemistry*, **461**, 27 (1984).
74. P.F. Bordui, R.G. Norwood, C.D. Bird and G.D. Calvert, *J. of Crystal Growth*, **113**, 61 (1991).

75. P.M. Bridenbaugh, J.R. Carruthers, J.M. Dziedzic and F.R. Nash, *Appl. Phys. Lett.*, **17**, 104 (1970).
76. G.G. Zhong and Z.K. Wu, in *International Quantum Electronics Conference*, (IEEE: 1980) Cat. # **80 CH 1561–0**, 631.
77. T.R. Volk, V.V. Krasnikov, V.I. Pryalkin and N.M. Rubinina, *Sov. J. Quantum Electron.*, **20**, 204 (1990).
78. J.K. Yamamoto, K. Kitamura, N. Iyi, K. S., Y. Furukawa and M. Sato, *Appl. Phys. Lett.*, **61**, 2156 (1992).
79. Y. Furukawa, A. Yokotani, T. Sasaki, H. Yoshida, K. Yoshida, F. Nitanda and M. Sato, *Appl. Phys. Lett.*, **69**, 3372 (1991).
80. R. Masse and J. Grenier, *Bull. Soc. fr. Mineral. Cristallogr.*, **94**, 437 (1971).
81. F.C. Zumsteg, J.D. Bierlein and T.E. Gier, *J. of Appl. Phys.*, **47**, 4980 (1976).
82. G. Gashurov and R.F. Belt in*Tunable Solid State Lasers for Remote Sensing*, R.L. Byer Gustafson, E.K., and Trebino, R., *Ed.*, 119 (Springer-Verlag: 1984).
83. P.F. Bordui, J.C. Jacco, G.M. Loiacono, R.A. Stolzenberger and J.J. Zola, *J. Crystal Growth*, **84**, 403 (1987).
84. L.K. Cheng and J.D. Bierlein, *Ferroelectrics*, **142**, 209 (1993).
85. J.D. Bierlein, A. Ferretti, L.H. Brixner and W.Y. Hsu, *Appl. Phys. Lett.*, **50**, 1216 (1987).
86. K. Buritskii, E.M. Dianov, A.B. Kiselev, V.A. Maslov and E.A. Shcherbakov, *Electronic Letters*, **27**, 1896 (1991).
87. D.K.T. Chu, J.D. Bierlein and R.G. Hunsperger, *IEEE Trans. on Utrason., Ferroelect. and Freq. Cont.*, **39**, 683 (1992).
88. H. Vanherzeele and J.D. Bierlein, *Opt. Letters*, **17**, 982 (1992).
89. N.K. Hansen, J. Protas and G. Marnier, *C.R. Acas. Sci. Paris*, **307**, 475 (1988).
90. G.D. Stucky, M.L.F. Phillips and T.E. Gier, *Chemistry of Materials*, **1**, 492 (1989).
91. T. Nishikawa, N. Uesugi and H. Ito, *Appl. Phys. Lett.*, **55**, 1943 (1989).
92. W. Wiechmann, S. Kubota, T. Fukui and H. Masuda, *Opt. Lett.*, **18**, 1208 (1993).
93. N.C. Wong, *Opt. Lett.*, **15**, 1129 (1990).
94. D. Lee and N.C. Wong, *J. of Opt. Soc. Am. B*, **10**, 1659 (1993).
95. D. Lee and N.C. Wong, *Opt. Lett.*, **17**, 13 (1992).
96. L.K. Cheng, L.T. Cheng, J.D. Bierlein, F.C. Zumsteg and A.A. Ballman, *Appl. Phys. Lett.*, **62**, 346 (1993).
97. L.T. Cheng, L.K. Cheng, J.D. Bierlein and F.C. Zumsteg, *Appl. Phys. Lett.*, **63**, 2618 (1993).
98. L.K. Cheng, J.D. Bierlein and A.A. Ballman, *J. Crystal Growth*, **110**, 697 (1991).
99. A.M. Clark in*Handbook of Lasers*, R. J. Pressley, *Ed.*, (CRC Press: Cleveland, Ohio, 1971) 3.
100. L.R. Marshall and A. Kaz, *J. of Opt. Soc. Am. B*, **10**, 1730 (1993).
101. L.T. Cheng, L.K. Cheng and J.D. Bierlein, in *OE'Lase*, Los Angeles, (SPIE: 1993) **1863**, 43.
102. A.F. Ferretti and T.E. Gier, *U.S. Patent #5,066,356*, 5 (1992).
103. R.A. Laudise, *U.S. Patent #4,654,111*, (1987).
104. W. Xing, H. Looser, H. Wüest and H. Arend, *J. of Crystal Growth*, **78**, 431 (1986).
105. T. Fukuda and Y. Uematsu, *Jap. J. of Appl. Phys.*, **11**, 163 (1972).
106. T. Varnhorst, O.F. Schirmer and H. Hesse, *J. of Crystal Growth*, **108**, 429 (1991).
107. G. Metrat and A. Deguin, *Ferroelectrics*, **13**, 527 (1976).
108. I. Biaggio, P. Kerkoc, L.S. Wu, P. Gunter and B. Zysset, *J. Opt. Soc. Am.*, **B9**, 507 (1992).
109. B. Zysset, I. Biaggio and P. Gunter, *J. Opt. Soc. Am.*, **B9**, 380 (1992).
110. W.R. Bosenberg and R.H. Jarman, *Opt. Lett.*, **18**, 1323 (1993).
111. W.J. Kozlovsky, W. Lenth, E.E. Latta, A. Moser and G.L. Bona, *Appl. Phys. Lett.*, **56**, 2291 (1990).
112. K. Kato, *IEEE J. of Quantum Electron.*, **QE-18**, 451 (1982).
113. P.N. Kean and G.J. Dixon, *Opt. Lett.*, **17**, 127 (1992).
114. W. Bosenberg and C.L. Tang, *Appl. Phys. Lett.*, **56**, 1819 (1990).

115. K.H. Hubner, *Neues Jahrb. Mineral. Monatsh.*, 335 (1969).
116. C.T. Chen, B. Wu, A. Jiang and G. You, *Scient. Sinica B*, **28**, 235 (1985).
117. R.K. Lu and C.T. Chen, *Acta Physica Sinica*, **34**, 827 (1985 (in Chinese)).
118. C.S. Willard and A.C. Albrecht, *Opt. Comm.*, **57**, 146 (1986).
119. A.J. Henderson, M. Ebrahimzadeh and M.H. Dunn, *J. Opt. Soc. of Am.*, **B7**, 1402 (1990).
120. R.C. Eckardt, H. Masuda, Y.X. Fan and R.L. Byer, *IEEE J. of Quantum Electronics*, **26**, 922 (1990).
121. L.K. Cheng, W. Bosenberg and C.L. Tang, *Appl. Phys. Lett.*, **53**, 175 (1989).
122. Y.X. Fan, R.C. Eckardt, R.L. Byer, J. Nolting and R. Wallenstein, *Appl. Phys. Lett.*, **53**, 2014 (1988).
123. W. Bosenberg, L.K. Cheng and C.L. Tang, *Appl. Phys. Lett.*, **54**, 13 (1989).
124. A. Jiang, F. Cheng, Q. Lin, Z. Cheng and Y. Zheng, *J. of Crystal Growth*, **79**, 963 (1986).
125. R.S. Feigelson, R.J. Raymaker and R.K. Route, *J. Crystal Growth*, **97**, 352 (1989).
126. L.K. Cheng, *Growth and Applications of Low Temperature Phase Barium Metaborate Crystals*, Cornell University, *Ph. D. Thesis*, (1988).
127. G. Dai, W. Lin, A. Zheng, Q. Huang and J. Liang, *J. Am. Ceram. Soc.*, **8**, (1990).
128. W.R. Bosenberg, R.J. Lane and C.L. Tang, *J. of Crystal Growth*, **108**, 394 (1991).
129. H. Kouta, Y. Kuwano, K. Ito and F. Marumo, *J. of Crystal Growth*, **114**, 676 (1991).
130. P.F. Bordui, G.D. Calvert and R. Blachman, *J. of Crystal Growth*, **129**, 371 (1993).
131. C.T. Chen, Y. Wu, A. Jiang, W. B., G. Tou, R. Li and S. Lin, *J. Opt. Soc. Am. B*, **6**, 616 (1989).
132. S. Zhao, C. Huang and H. Zhang, *J. of Crystal Growth*, **99**, 805 (1990).
133. S.P. Velsko, M. Webb, L. Davis and C. Huang, *IEEE J. of Quantum Electron.*, **27**, 2182 (1991).
134. D.W. Chen and J.T. Lin, *IEEE J. of Quantum Electron.*, **QE-29**, 307 (1993).
135. Y. Wang, Z. Xu, D. Deng and W. Zheng, *Appl. Phys. Lett.*, **59**, 531 (1991).
136. Y. Tang and Y. Cui, *Opt. Lett.*, **17**, 192 (1992).
137. G. Robertson, A. Henderson and M.H. Dunn, *Appl. Phys. Lett.*, **60**, 271 (1992).
138. T. Ukachi, R.J. Lane, W.R. Bosenberg and C.L. Tang, *Appl. Phys. Lett.*, **57**, 980 (1990).
139. G. Robertson, A. Henderson and M. Dunn, *Opt. Lett.*, **16**, 1584 (1991).
140. D. Withers, G. Robertson, A.J. Henderson, Y. Tang, Y. Cui, W. Sibbett, B.D. Sinclair and M.H. Dunn, *J. of Opt. Soc. Am. B*, **10**, 1737 (1993).
141. M. Ebrahimzadeh, G.J. Hall and A.I. Ferguson, *Appl. Phys. Lett.*, **60**, 1421 (1992).
142. M. Ebrahimzadeh, G.J. Hall and A.I. Ferguson, *Opt. Lett.*, **17**, 652 (1992).
143. H. Zhou, J. Zhang, T. Chen, C. Chen and Y.R. Shen, *Appl. Phys. Lett.*, **62**, 1457 (1993).
144. Y. Cui, M.H. Dunn, C.J. Norrie, W. Sibbett, B.D. Sinclain, Y. Tang and J.A.C. Terry, *Opt. Lett.*, **17**, 646 (1992).
145. M.P. Scripsick, X.H. Fang, G.J. Edwards, L.E. Halliburton and J.K. Tyminski, *J. Appl. Phys.*, **73**, 1114 (1993).
146. J.P. Chernoch, M.J. Kukia, W.T. Lotshaw and J.R. Unternahrer, in *Topical Meeting on Advanced Solid-State Lasers*, Santa Fe, New Mexico, (Optical Society of America: 1992) **13**, 386.
147. S.T. Yang, C.C. Pohalski, E.K. Gustafson, R.L. Byer, R.S. Feigelson, R.J. Raymakers and R.K. Route, *Opt. Lett.*, **16**, 1493 (1991).
148. J.Y. Huang, Y.R. Shen, C.T. Chen and B. Wu, *Appl. Phys. Lett.*, **58**, 1579 (1991).
149. H.J. Krause and W. Daum, *Appl. Phys. Lett.*, **60**, 2180 (1992).
150. G. Robertson and M.H. Dunn, *Appl. Phys. Lett.*, **62**, 3405 (1993).
151. A. Del Corno, G. Gabetta, G.C. Reali, V. Kubecek and J. Marek, *Opt. Commun.*, **91**, 93 (1990).
152. V. Kubecek, Y. Takagi, K. Yoshihara and G.C. Reali, *Opt. Commun.*, **91**, (1992).

# 6. EXAMPLES OF PRACTICAL OPTICAL PARAMETRIC OSCILLATORS AND AMPLIFIERS

Because of the limited availability of suitable combinations of pump sources and nonlinear crystals in the early days, the development of practical optical parametric devices was initially focused on singly-resonant nanosecond-type of oscillators pumped by $Q$-switched solid-state lasers. Subsequent developments have been extended to include broadly tunable picosecond and femtosecond optical parametric oscillators and amplifiers. Most recently, there has also been significant progress made on cavity-length-stabilized doubly-resonant cw OPO's.[1,2,3]

Practical parametric generators are either oscillators or parametric amplifiers seeded by spontaneous parametric emissions depending on available pump sources and nonlinear optical crystals. All such devices are currently limited by the same "average-power barrier" on the order of a few watts. Because of the higher conversion efficiency possible, it is always preferable to have an oscillator if it can be achieved. On the other hand, if the pulse energy is high, the duty cycle must be low. At low duty cycles, it is not possible to achieve oscillation with synchronous pumping because of cavity-length limitations and the device must be an amplifier.

In the nanosecond time domain, pump sources derived from $Q$-switched lasers, such as Nd:YAG lasers and its harmonics, can easily have energies in the 10 mJ to J/pulse range and intensities far greater than 100 MW/cm$^2$ with pulse lengths on the order of 8–10 nsec. At such pump power levels, the total single-pass gain in, for example, a 1-cm long BBO crystal, would be over 50% and the pump pulse duration is long enough for 15–20 passes in a typical 15 cm long cavity. The corresponding multi-pass gain is, therefore, sufficient to amplify the spontaneous parametric emission to the near saturation level of the optical parametric oscillator. Thus, in the nanosecond time domain, quasi-cw pumped optical parametric *oscillators* with efficiency in the 30–50% range could be built.

For picosecond pulses, the limited pump pulse duration obviously does not permit quasi-cw pumping. The parametric generator is generally pumped in a traveling-wave mode in which the pump pulse travels in synchronism with the signal pulse to provide parametric gain in the nonlinear crystal. At low pump cycles, picosecond pulse energies on the order of 5 $\mu$J or more and intensities in the 10 GW/cm$^2$ or greater range are possible. At such intensity levels, the net gain is more than 50 dB in a 1-cm long BBO crystal, for example. Broadly tunable picosecond generators with net output in the $\mu$J/pulse levels consisting of single-pass optical parametric power amplifiers (OPA) seeded by amplified spontaneous parametric emission have been demonstrated.[4,5,6] These sources are, however, generally limited to repetition rates of 100 KHz or lower so that the average power is on the Watt-level. Because of the high peak powers, such sources are particularly useful for nonlinear optical applications in the picosecond time domain.

In the femtosecond time domain, the nonlinear optical crystal for parametric conversion must be very thin (on the order of a mm) to avoid pulse broadening due to group velocity dispersion. On the other hand, CW mode-locked pump sources over one watt, such as the Ti-sapphire laser, are now available. The corresponding peak power can be over 100 KW at 100 MHz with peak intensities greater than 1 $GW/cm^2$ in a suitably focused beam. With highly nonlinear crystals such as KTP, single-pass gain of a few % can be readily achieved even for very thin parametric-conversion crystals. With synchronous pumping at such a high repetition rate (100 MHz or higher), it is, therefore, possible to sustain truly broadly tunable CW femtosecond optical parametric *oscillations* at nJ/pulse levels. Extension of this approach to the picosecond time domain has so far been limited to the generation of 1–10 psec pulses at similar energy levels.[7,8] Due to the higher duty cycles involved, to achieve comparable cw optical parametric oscillation in the psec time domain at multi-$\mu$J/pulse levels would require hundreds-to-kilo watts of cw pump power, which is not available.

Examples of different types practical optical parametric generators are reviewed in this chapter in some detail.

## 6.1 HIGH-ENERGY NANOSECOND PARAMETRIC OSCILLATORS

With a $Q$-switched laser as the pump, the pulse duration of the OPO output is generally in the 5 to 10 nanosecond range and the pulse repetition rate is typically on the order of 10–30 Hz. Practical OPO's with outputs in the 10 mJ to J per pulse range are now possible.[9–11] Because the OPO is scalable, the output energy or power of the OPO is, in principle, only limited by that of the available pump source. The energy efficiency of such devices can also be very high, typically in the 30% to 60% range.

Earlier development of such OPO's tended to be in the infrared using mainly $LiNbO_3$ and $LiIO_3$ crystals.[12,13] The advantage of these crystals is that they have relatively high effective nonlinear optical coefficients. The problem with $LiNbO_3$ was that earlier crystals had a relatively low optical damage threshold due to the photo-refractive effect, [i.e. the index of refraction of $LiNbO_3$ would change when exposed to even moderately intense visible light over a relatively short period of time], although the improved quality of recent $LiNbO_3$ crystals has led to significantly higher damage thresholds and some practical OPO applications. $LiIO_3$ is highly hygroscopic and, thus, impractical for broad applications. In addition, with the uv absorption edge of these materials in the 400 nm range, pumping with the Nd:YAG laser is limited to its fundamental at 1.06 $\mu$m or the second-harmonic at 533 nm and, hence, the tuning range of such OPO's is limited to the visible or longer wavelength range. Recent advances in truly broadly tunable practical OPO's resulted from the development of nonlinear optical crystals that have high damage thresholds, are transparent and phase-matchable well into the uv, and have other properties uniquely suited for OPO applications as discussed in Chapter 5.

### 6.1.1 $\beta$-Barium Borate (BBO) OPO

$\beta$-barium borate (BBO) is one of the first nonlinear crystals to meet all the criteria for broad based OPO applications. Although growth of large BBO crystals of good optical quality was initially difficult, recent developments of growth techniques such as the immersion-seeded technique or the Czochralski method of growing large barium borate crystals has substantially solved the problem making commercialization of BBO OPO's possible.

As discussed in the previous chapter, the uv absorption edge of BBO is near 200 nm. For OPO applications, it can be pumped at the third-harmonic of Nd:YAG at 355 nm. The corresponding tuning range of the signal and idler ouputs extend from 400 nm to 2.5 $\mu$m [Figure 2.3]. In fact, this entire tuning range can be covered with a single set of mirrors resonating the signal branch in the visible. With the addition of an efficient BBO frequency-doubler, continuously tunable radiation from 200 nm to 2.5 $\mu$m is now available from a single BBO OPO system.

BBO OPO's pumped by the second,[14] third,[14,15] or fourth[16] harmonic output of the Nd:YAG laser have been reported extensively in the literature. The more optimal choice at this time is pumping by the third-harmonic at 355 nm using either Type I or Type II interaction. It gives a broad tuning range (from approximately 400 nm to 2.5 $\mu$m) and yet the mirror uv-damage problem is still manageable at this pump wavelength. With a fourth-harmonic pump at 266 nm, a tuning range from 300 nm to 2.5 $\mu$m (Figure 2.3) is possible using multiple sets of mirrors, but the uv mirror-coating damage problem becomes severe. The tuning range with second-harmonic pumping is more limited.

Figure 4.1 shows the schematic of a third-harmonic of Nd:YAG laser pumped BBO OPO operating in the nano-second range. This OPO embodies two design features that might be useful for other types of OPO's as well. The first feature is the two-crystal walk-off compensation arrangement. As shown in Figure 4.1, because the Poynting vector and the wave vector of the extraordinary pump beam are not in the same direction while those for the ordinary signal beam are in the same phase-matched direction, the pump beam walks away from the signal beam. In the case of BBO pumped at the third harmonic of YAG, the walk-off length is typically about 1 cm at the wavelengths of interest which is the effective limit of the interaction length. One way to compensate for this walk-off effect and increase the interaction length is to use more than one crystal with their $c$-axes alternately oriented in such a way that the pump beam walks away from the signal beam in one crystal and walks back onto it in the following crystal as shown in Figure 4.1. A significantly lower threshold and higher efficiency could be achieved by using multiple crystals with such a walkoff-compensation scheme.

The second useful design feature is the use of the beam-steering mirrors to couple the pump beam into and out of the OPO cavity. The purpose of these mirrors is to avoid uv damage of the mirror coatings at the pump wavelength. Because the damage threshold of uv-transmitting and visible-reflecting mirror coatings is generally lower than that of uv-reflecting and visible-transmitting mirror coatings, the use of beam-steering mirrors substantially increases the pump power level that can be tolerated in practical uv-pumped OPO's. As the energy level of third-harmonic YAG laser pumped BBO OPO reaches the hundreds of mJ level, mirror damage is an increasingly important design consideration. The beam-steering scheme also allows pumping at 266 nm leading to parametric oscillation in BBO down to 300 nm.

The linewidth of the nanosecond OPO's is an important design issue. For a Type I phase-matched BBO OPO pumped at 355 nm, without the use of any special line-narrowing scheme, the linewidth can vary from a few Å far from the degenerate point to nearly 100 Å near degeneracy. Oscillator linewidth of this order of magnitude may be adequate for some applications, but for other high-resolution spectroscopic applications, it is important to reduce the linewidth. As discussed in Section 4.4, the dominant line broadening mechanisms are due to pump beam linewidth and finite pump beam divergence. At the degenerate point, the signal and idler wavelengths are the same and the wavelength versus phase-matching angle curve is vertical for Type I action. Near the degenerate point, a small pump beam divergence can, therefore, lead to a very large parametric linewidth. To reduce the linewidth, in general, additional line-narrowing elements must be introduced into the oscillator cavity.[17,18] In the case of BBO OPO, the use of a grating in Littrow-configuration replacing one of the OPO mirrors typically can reduce the line-width down to 2 or 3 Å throughout the tuning range. Using the grating in the Littman-configuration can further reduce the tuning range to approximately 0.3 Å.[16] In both cases, the trade-off is that the threshold for oscillation is typically raised by 50 to 200% with a corresponding reduction in the output power and efficiency. To obtain narrow linewidth at high output power levels, another commonly used approach is to injection-lock a high power oscillator with a low-power narrow-linewidth oscillator. Injection-locking the pump laser to achieve narrow pump linewidth and better collimation will further improve the OPO linewidth. The use of a combination of these schemes can lead to single longitudinal mode operation of practical BBO OPO's at power levels up to 100 mJ.

The use of Type II interaction can also lead to narrower linewidth[19] because the signal and idler waves have different polarizations and, hence, the tuning curves of these waves generally cross at an angle where the signal and idler wavelengths are the same. There is, therefore, no true degenerate point where the signal and idler wavelengths and the dispersions are the same. With Type II phase-matched

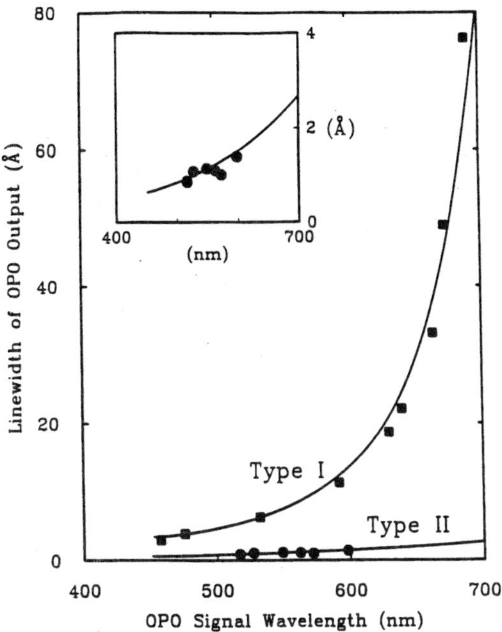


**Fig. 6.1** The linewidth vs. tuning for the Type I (squares) and Type II (dots) 355 nm pumped BBO OPOs. The solid lines represent calculated linewidths. The Type II OPO has significantly narrower linewidth throughout the tuning range. The inset shows the Type II data with an expanded ordinated scale. (Figure 1, Ref. 19).

BBO OPO pumped at 355 nm, it is possible to maintain an oscillator linewidth less than 2 Å over most of the tuning range (Figure 6.1) without the use of grating reflectors. It should be remembered that in general the oscillator linewidth is determined by the slope of the tuning curve, i.e. wavelength versus phase-matching angle, along with the divergence of the pump beam. Depending on the crystal dispersion, vertical slope in the tuning curve can even exist for Type II processes not at degenerate points.[20] This is particularly true when the lower energy idler wave approaches the ir cutoff. Examples of crystals exhibiting such tuning characteristics include LBO[21] and the KTP isomorphs (see Figures 6.10–6.13).

For high power OPO's, it is often difficult to achieve single mode oscillation even using the various line narrowing schemes discussed above. A possible alternative is to use a single-mode low-power OPO to injection-lock a high-power OPO.[22] This approach has been successfully demonstrated in commercial BBO and KTP OPO's.[22,23] In scaling to higher average power, absorption induced thermal loading of the nonlinear crystal becomes another important issue. It is reported[24] that thermally induced dephasing problem in a 308 nm laser pumped BBO OPO's can become important at an average power as low as ~10 W, and proper thermal management would be needed to achieve optimal performance for these high average power OPO's.

With the successful operation of the BBO OPO, many other choices of pump sources and nonlinear crystals have followed and still more are to be evaluated and explored. For example, in the visible and near infrared region, OPO's based on the material LBO have been demonstrated.[25] In addition to improving the performance of the OPO's in the spectral range already covered, extending the tuning range is clearly an important direction of future research and development. With crystals such as some of the KTP isomorphs, $KNbO_3$, and ternary semiconductors such as $ZnGeP_2$, sources for the important 3 to 5 $\mu$m range are now being actively developed.[26–29] The use of materials such as $AgGaS_2$ and $AgGaSe_2$ could in principle extend the OPO tuning range to 20 $\mu$m, but the growth technology of these materials still needs significant improvement. Beyond 20 $\mu$m, it will be increasingly more difficult to find suitable nonlinear crystals that are still transparent.

## 6.2 CONTINUOUS-PULSE-TRAIN, HIGH-REPETITION RATE, FEMTOSECOND PARAMETRIC OSCILLATORS

Although $Q$-switched laser pumped nanosecond type of OPO's dominated the early development, recent efforts on optical parametric oscillators have extended down to the femtosecond time domain.

Until recently, almost all the experiments on ultrafast processes in the femtosecond time domain had made use of sources derived in one form or another from the Rh6G dye laser operating at approximately 630 nm with an average power in the range of tens of mW. There was a need to extend the wavelength range and increase the power of the femtosecond sources while maintaining the high repetition rate of the Rh6G laser.

Broadly tunable high repetition-rate femtosecond sources are particularly important for studying ultrafast processes, because broad tunability allows a greater variety of materials and processes to be studied and because of the higher signal-to-noise ratio possible in such experiments when the repetition rate is high. There have been dramatic advances in the development of such sources in the last few years following the first demonstration of a broadly tunable femtosecond optical parametric oscillator (fs OPO).[30]

In the development of tunable femtosecond sources, to achieve broad tunability without sacrificing time resolution has been a well recognized goal. The need to achieve these at sufficiently high pulse repetition rate has not been emphasized, however. Yet, in the study of ultrafast relaxation processes, as the time resolution increases, more details of the relaxation dynamics will inevitably be revealed, including, for example, the possibility of multiple or stretched exponential processes. To analyze the data from the usual pump-probe type of experiments in such studies, more accurate data and sophisticated methods are needed beyond simple graphic methods such as drawing asymptotes to the decay curve, which is notoriously inadequate when more than one or two simple exponentials are involved. Higher pulse repetition rate leads to higher signal-to-noise ratio and, hence, more accurate data. Thus, there is a need to achieve broad tunability, short pulse width, and sufficiently high repetition rate ($\sim$100 MHz or higher) at the same time. *These requirements can be met simultaneously at reasonable power levels only in a femtosecond optical parametric oscillator (fs OPO)*, because of the following reasons: To achieve truly broad continuous tunability, the device must be some sort of parametric device. To maintain the shortest possible pulse width, the parametric conversion crystal must be very thin to avoid group velocity dispersion and to allow phase-matching of the entire femtosecond spectrum. For a thin parametric crystal, the single pass conversion efficiency will typically be relatively low at currently available pump power levels because of the lack of high power femtosecond amplifiers at high repetition rates. Thus, the parametric device must be an oscillator, not a single or double pass parametric amplifier. For optical parametric oscillators, the conversion efficiency can be in the range of 30 to 50% at a few times above the oscillation threshold, although it may take thousands of round trips in the cavity to build up from the spontaneous emission level to the steady-state oscillator output level.

The first broadly tunable high repetition rate femtosecond source was a KTP OPO[30] synchronously pumped inside the cavity of a Rh6G dye laser. The average output power level in the visible was only a few mW with pulse widths on the order of 100 fs at $10^8$ Hz rate. The tuning range was, however, very large, from 700 nm to nearly 3.5 $\mu$m.[31] Some experiments were done using this type of intra-cavity pumped KTP OPO as the source, but its practical utility was limited because aligning and tuning such a device was difficult in practice. These problems were, however, not intrinsic to the femtosecond OPO. It was mainly due to the fact that the threshold for oscillation of the KTP femtosecond OPO could not be reached easily with the Rh6G femtosecond dye lasers available at the time. This required the OPO to be pumped inside the pump laser cavity. The coupling of the pump laser cavity with the OPO cavity made it difficult to align and tune the laser. The situation has since changed dramatically with the development of the Ti-doped sapphire laser[32] which can generate over 2 W of mode-locked power at $\sim$100 MHz with pulse widths under 100 fs. The increased pump power makes it possible to pump the KTP femtosecond OPO external to the pump laser cavity. Thus, with the pump laser cavity and the OPO cavity decoupled and separately alignable, all the initial difficulties associated with the KTP femtosecond OPO are reduced. Ti:sapphire laser pumped fs OPO's as a tunable femtosecond source is now a practical and commercially available device.

As a comparison, in the nanosecond time domain, the pulse length of typical pump lasers on the order of 5–10 nanoseconds is long enough to allow multiple round-trip passes in the cavity for an OPO to reach near the saturation level of an oscillator. In contrast, the pump pulse length for fs OPO's is typically on the order of tens of femtoseconds and the corresponding spatial extent is on the order of tens of microns which is much shorter than the OPO cavity length. Also, unlike in lasers, there is no energy-storage in the nonlinear crystal in the femtosecond OPO. The pump, signal, and idler pulses must overlap in the nonlinear crystal which means that the pump pulse train must be in a traveling-wave mode and in synchronism with the signal pulse as it traverses the OPO cavity. Maintaining synchronism for short pulses in a very thin crystal over many round-trips in the cavity is extremely difficult. For pulses on the order of 100 fs or less, the OPO cavity length must be controlled to the order of 100 nm or less. The design considerations and the characteristics of the low gain cw femtosecond OPO's are, therefore, very different from those of the nanosecond type of OPO's discussed in the previous section and the picosecond type of optical parametric amplifiers discussed in the following section. The basics of the fs OPO's are discussed in this section. The dye laser intracavity pumped case will be discussed first. The more recent results on Ti-sapphire laser external-cavity pumped OPO's will follow.

### 6.2.1 Rh6G Dye Laser Pumped Femtosecond KTP OPO

When the first experiment on the fs OPO was carried out,[30] the Rh6G-DODCI mode-locked dye laser was the only available pump in the femtosecond time domain. The average pump power was on the order of a few tens of mW with peak powers on the order of a few KW. Taking into consideration the nonlinear coefficient, group velocity dispersion, damage threshold, etc., the optimal choice of the nonlinear optical crystal for OPO applications in the sub-100 fs time domain was thought to be KTP. For Type II parametric interaction ($o \rightarrow e+o$) in the biaxial KTP crystal, for example, the corresponding group velocity dispersion $\Delta v_g^{-1}$ is 53 fs/mm, at a pump wavelength of 630 nm and signal wavelength of 850 nm. This implies that the length of the nonlinear crystal should be on the order of a mm. Even though KTP is a material with a relatively large effective nonlinear coefficient, the effective gain coefficient at a few KW peak pump power focused down to a spot of diameter on the order of 100 $\mu$m external to the dye laser cavity is still too small to reach the threshold for optical parametric oscillation. The only possibility is to place the nonlinear crystal inside the Rh6G dye laser cavity to take advantage of its higher intracavity power to pump the OPO. Peak intensity in excess of $10^9$ W/cm$^2$ are possible at an intracavity focus of such a dye laser.

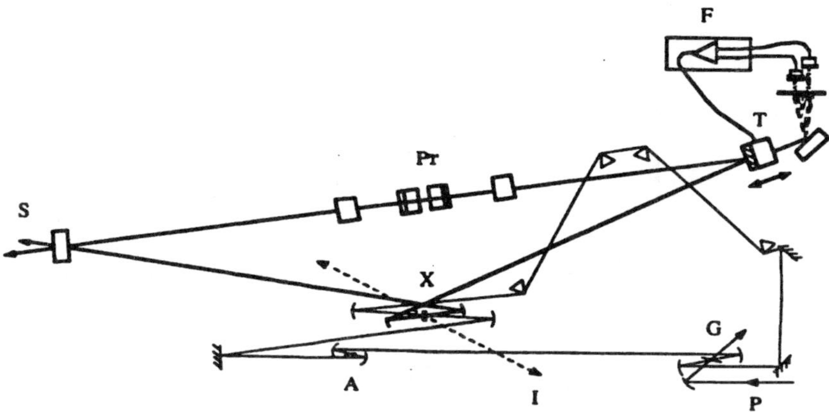


**Fig. 6.2** Schematic of femtosecond optical parametric oscillator. The beamline for the signal (S) is shown in thin solid line. The dotted lines are idler beams (I) which exit the crystal (X) as shown. One flat mirror is mounted an a piezoelectric transducer (T) for length adjustment. This is driven by feedback loop (F) to maintain proper synchronism. Dispersion compensation is provided by four prisms (Pr) in the vertical plane. The oscillator is intracavity pumped by a CPM dye laser. P, G, and A are the Ar+ laser pump, the gain jet and the absorber jet, respectively.

With the nonlinear crystal inside the mode-locked femtosecond pump laser cavity, the cavities of the pump laser and the OPO become closely coupled so that adjustment of one cavity often affects the alignment of the other. This makes the intracavity pumped OPO exceedingly difficult to operate. The configuration of the first operational intracavity-pumped dispersion-compensated KTP fs OPO is shown in Figure 6.2 (Edelstein and Wachman theses, Ref. 33). The KTP crystal is 1.4 mm long. The pump is a standard Rh6G-DODCI colliding-pulse mode-locked (CPM) laser with the following modifications: the output coupler is replaced by a high reflector, a third intracavity focus is formed by two $r = 20$ cm high reflectors, and the dispersion-compensation quartz prisms pairs are separated by an additional 14 cm to compensate for the group velocity dispersion in 1.4 mm of KTP at 620 nm ($d^2n/d\lambda^2 = 0.723\ \mu\text{m}^{-2}$).

The calculated tuning curve of the KTP OPO is shown in Fig. 6.3. To reach the largest tuning range, the OPO is designed to operate below phase-matching angle of about 65°. Since the initial alignment is based upon the very weak spontaneous parametric emission, for ease of alignment, the shorter wave length $e$-wave with better detectors is resonated and is the signal wave. Also, because the pump and signal waves are polarized in orthogonal directions (the pump being an $o$-wave and the signal being an $e$-wave), the arrangements of the dispersion-compensation prisms in the pump dye

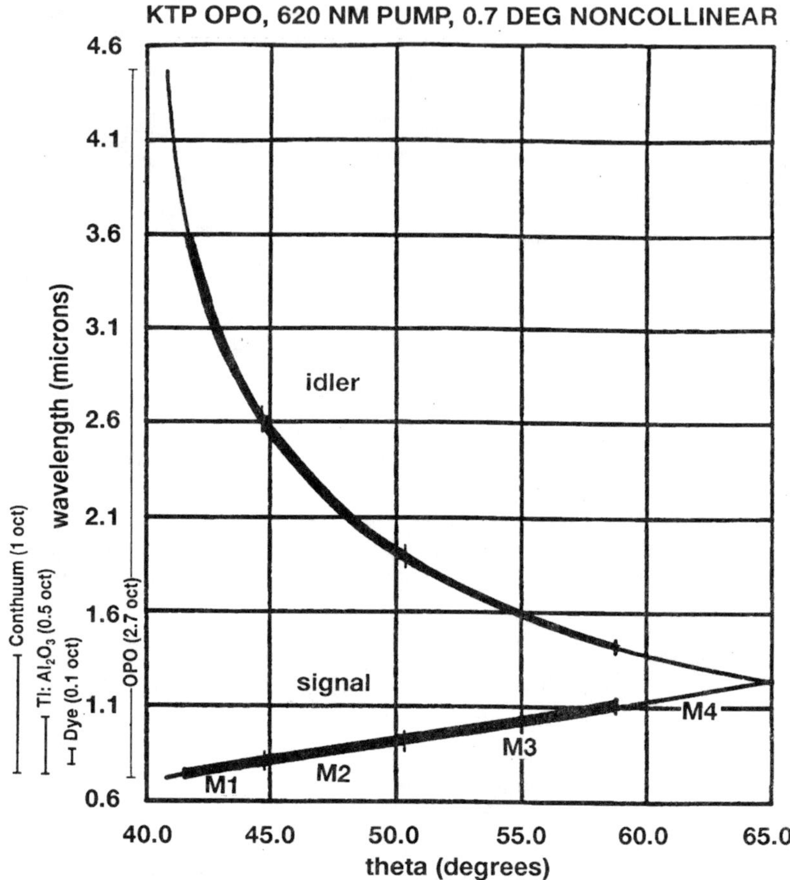


**Fig. 6.3** Tuning curve for the femtosecond OPO with demonstrated tuning (bold ink) and mirror sets (Mi) to get the full range (720±4500 nm). This represents 2.6 octaves (oct.) in the optical frequency. Comparisons are made with other femtosecond pulsed light sources.

laser cavity and that of the OPO must be different, with the displacements in one cavity in the horizontal plane and the other in the vertical plane as shown in Figure 6.4 (Edelstein and Wachman theses, Ref. 33).

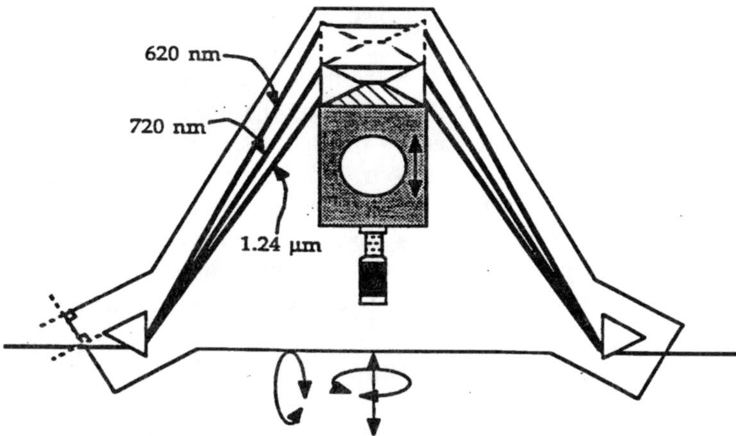


**Fig. 6.4** Modular prism interferometer. Brewster prisms are oriented with respect to carefully machined edges at the appropriate angles, and all distances are defined by machining, thus insuring symmetry (asymmetry leads to chromatic aberration). The upper two prisms are mounted on a single translator so they may easily be adjusted to intersect all relevant beams (620–1240 nm). The baseplate is mounted on a tilting, vertical translating base to allow alignment of the unit as a whole, and adjustment of lower prism glass in the beam.

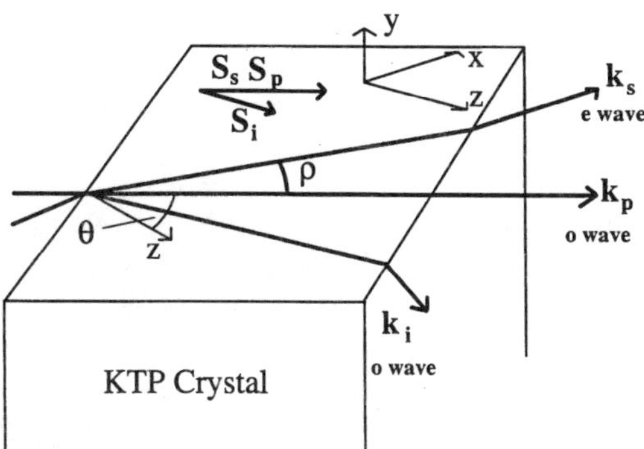


**Fig. 6.5** Orientation of KTP crystal and the direction of k-vectors and Poynting vectors (S) of the pump, signal, and idler waves in a femtosecond OPO. Note that, to compensate for walkoff, the Poynting vectors of the pump and signal waves are collinear in a noncollinear Type II phase-matching configuration.

The crystal is a hydrothermally grown KTP cut for Type II ($o \rightarrow e + o$) phase matching at near normal incidence with a single layer of $MgF_2$ antireflection coating on each surface. Hydrothermally grown KTP crystals appear to have a higher optical damage threshold than the flux (phosphate or tungstate) grown types of KTP. The orientations of the crystal and the directions of polarizations and the k-vectors are shown in Figure 6.5. The corresponding tuning curve is shown in Figure 6.3 (Edelstein and Wachman theses, Ref. 33).

The crystal is oriented in such a way to maximize the effective coefficient, $d_{\text{eff}}$, and achieve the desired tuning characteristics. For small non-collinear phase-matching angles ($\rho$), Eqs. (5.14) and (5.15) gives a good approximation of $d_{\text{eff}}$ for SHG for KTP:

$$\text{Type I} \qquad d_{\text{eff}} = (1/2)[d_{15} - d_{24}] \sin(2\theta) \cos(2\phi) \tag{5.14}$$

$$\text{Type II} \qquad d_{\text{eff}} = -[d_{15} \sin^2\phi + d_{24} \cos^2\phi] \sin(\theta) \tag{5.15}$$

where $\theta$ and $\phi$ are defined in Figure 6.5. For KTP, the values of $d_{15}$ and $d_{24}$ are $\sim$1.9 and 3.6 pm/V, respectively. For these values, the Type II interaction always has a larger $d_{\text{eff}}$ than does Type I interaction. For Type II interaction, $d_{\text{eff}}$ is maximized for $\phi = 0$ so that the KTP crystal is aligned such that the parametric interactions occur in the x–z plane. It is clear from Eq. (5.15) that one maximizes $d_{\text{eff}}$ for the Type II interaction for a phase-matched angle of $\theta = 90°$. The value of $\theta$ is, however, usually dictated by the tuning characteristics.

Because of better detector sensitivity, the preference is to resonate the wave below 1 $\mu$m. As can be seen from the tuning curves, the short wavelength branch below 1 $\mu$m is an extraordinary wave in the phase-matching region with the largest total tuning range. With a resonated extraordinary signal wave and an ordinary pump wave in a Type II interaction, and to minimize the Fresnel losses of the nonlinear crystal for both waves, antireflection coating of the crystal at normal incidence must be used instead of cutting the crystal at Brewster angle.

With orthogonally polarized resonated extraordinary signal and ordinary pump waves, the corresponding Poynting vectors are not collinear with collinear phase-matching. The two waves will, thus, walk away from each other. It was found empirically that when curved focusing mirrors are used as shown in Figure 6.2 the OPO optimizes itself by choosing the interaction to be in such way that the Poynting vectors of the pump and signal waves are in the same direction, thus, ensuring maximum overlap of these two beams[30,33] (Figure 6.6). This leads to a slightly noncollinear phase-matching angle between the pump and signal wave vectors of 2.8° in the crystal.

This fact[31,34] has interesting ramifications for the design of practical singly resonant OPO's quite generally. Normally, to minimize walkoff in anisotropic crystals, one resonates the parametrically generated wave of the same polarization as the pump in a collinear Type II phase-matching geometry. This results in collinear Poynting vectors

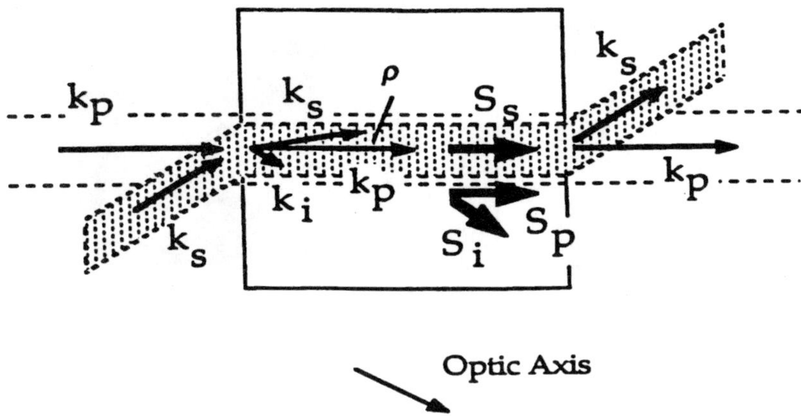


**Fig. 6.6** Walkoff compensation in the fs OPO. The signal is aligned relative to the pump so that the angle between $k_s$ and $k_p$ in the crystal is equal to the negative of the walkoff angle, $\rho$; the Poynting vectors of the signal and pump, $S_s$ and $S_p$, are therefore parallel as the beams traverse the crystal, though this occurs at the expense of the nonresonant idler ($k_i$ and $S_i$).

as well as the $k$-vectors of the pump and resonated waves. The above observation shows that what is important is the collinearity of the Poynting vectors, as long as the noncollinear angles between the wave vectors of the interacting waves are not too large so that the corresponding interaction length becomes shorter than the physical length of the crystal. This means that it is possible to resonate the wave with orthogonal polarization to the pump in a noncollinear phasematching geometry (the internal noncollinear angle equal to the walkoff angle) as shown in Figure 6.6. Since this conclusion is equally applicable to both Type I and Type II configurations and for signal waves of either polarization, it increases somewhat the range of choices available for the design of singly resonant OPO's operating in any time domain. It should be noted, however, that when the noncollinear angle becomes too large, eventually the effective interaction length is going to be limited by the walk-off angle of the idler wave as well as that of the signal wave which in turn will raise the oscillation threshold. Walk-off of the idler wave will tend to make the parametric gain to increase linearly rather than as $\sim \cosh\ gz$ as given in Eq. (2.29).

For the intracavity pumped OPO, the initial alignment of the device is exceedingly difficult due to the number of intracavity optical elements and the fact that the pump laser cavity and the OPO cavity are directly coupled. The key to the alignment procedure is to make sure that all the elements are first positioned and oriented to very close tolerances so that it will be possible to detect the multi-pass spontaneous parametric emission and optimize its amplification to achieve oscillation.

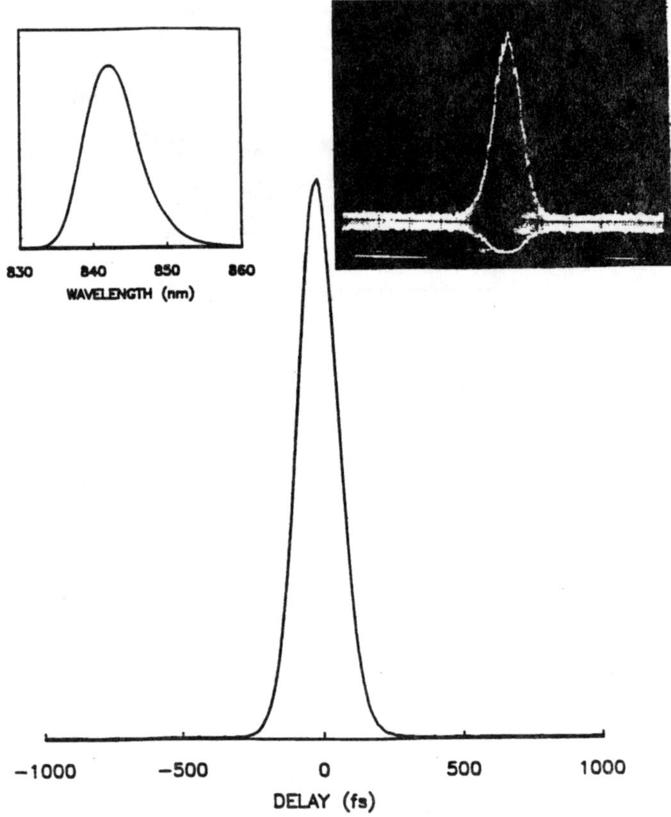


**Fig. 6.7** Autocorrelation trace for the signal at 840 nm in the dispersion compensated OPO with near-perfect fit to $\mathrm{sech}^2$. The spectrum is shown in the inset on the left. It is smooth and highly symmetric. The pulse is transform limited ($\Delta\nu\Delta\tau = 0.35$) and has a width of 105 fs. The interferometric autocorrelation of the pulses is shown in the inset on the right. It shows the proper contrast ratio, flat background, and clean structureless wings. these are indicative of stable transform-limited modelocked operation.

The intracavity pumped KTP OPO was the first broadly tunable femtosecond source to operate. Tuning was demonstrated from approximately 700 nm to 3.5 $\mu$m (Figure 6.3).

Figure 6.7 displays an intensity autocorrelation (AC) trace of the signal pulses with the dispersion-compensation prisms adjusted for near-zero net intracavity group velocity dispersion (GVD) at 840 nm. An interferometric AC and spectrum of the pulse are shown in the inset of Figure 6.7. The signal pulse width of 105 fsec (a $\mathrm{sech}^2$ fit) was one half that in the absence of the prisms and the pump pulse width was 170 fs.

The signal pulse time-bandwidth product $\Delta\nu\Delta\tau$ of $\approx$0.35 is indicative of transform-limited pulses and is also substantiated by the highly symmetric spectrum and the structureless wings of the AC's. Significantly, this minimum OPO pulse width could be obtained with pump pulses at least as long as 350 fsec, suggesting that the signal pulse width is limited primarily by group delay between the signal and pump pulses in the crystal. Careful compensation of the group velocity dispersion should result in much shorter signal pulse widths. The average output power level in the visible was typically 2 mW per arm. The tuning range was, however, very large, from 720 nm to over 3.5 $\mu$m. Although some experiments were carried out using this intra-cavity pumped KTP OPO as the source, aligning and tuning such an intracavity-pumped device was difficult in practice.

### 6.2.2 Ti:Sapphire Laser Pumped Femtosecond KTP OPO

A major breakthrough came with the development of the cw mode-locked titanium-doped sapphire laser. The average power level of the Ti-sapphire laser could be in the 2 to 3 W range with pulse widths less than 100 fs at $10^8$ Hz repetition rate. Since the average pump power required to achieve oscillation in an externally pumped angle-tuned KTP fs OPO is on the order of 100 mW, it can be exceeded easily by the Ti-sapphire fs laser. External pumping decouples the fs laser cavity and the fs OPO cavity, allowing independent adjustment and optimization of each.

Operation of Ti-sapphire laser pumped sub-100 fs KTP OPO's was first reported in 1992.[35,36] This was quickly followed by fs optical parametric operations in lithium tri-borate[37] (LBO, $LiB_3O_5$), potassium titanyl arsenate[38] (KTA, $KTiOAsO_4$), cesium titanyl arsenate[39] (CTA, $CsTiOAsO_4$), and rubidium titanyl arsenate[40] (RTA, $RbTiOAsO_4$), potassium niobate ($KNbO_3$) etc.

Figure 6.8 shows the schematic of the first Ti-sapphire laser pumped KTP OPO with intracavity dispersion-compensation prisms as reported in Ref. 35. These prisms are important. Without them (Ref. 36), the pulse width will be wide and there will be substantial chirping in the wings out to hundreds of femtoseconds even with extracavity pulse compression and dispersion compensation to reduce the chirp, even though the width of the coherent peak part of the pulse is reported to be as short as 62 fs. With the intracavity prisms, ideal chirp-free pulses down to 57 fs have been obtained (Figure 6.9). Because the parametric process is an elementary three-photon process, there is no energy storage in the cavity and the pulse width of the signal is primarily determined by that of the pump. Potentially, the pulse width can be down to the 20 fs range with sufficiently short pump pulses and appropriate compensation for the group velocity mismatch. The phase-matching bandwidth at, for example, a signal wavelength of 1.3 $\mu$m and a pump wavelength of 800 nm is on the order of 280 nm-mm for KTP and KTA and a little less for CTA, allowing ultra-short pulse operation.

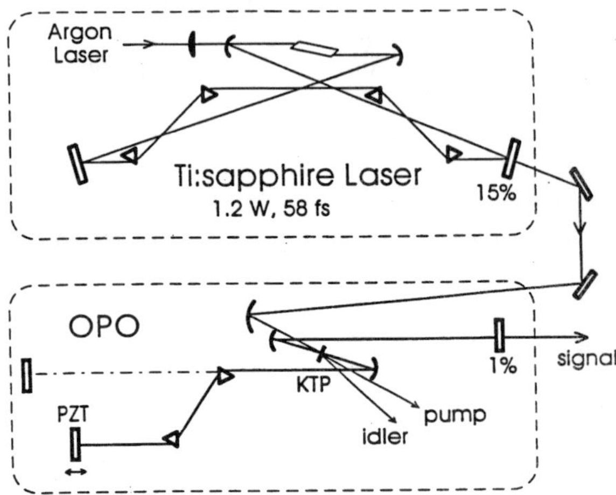


**Fig. 6.8** Schematic of ring-cavity Ti:sapphire laser pumped broadly tunable high-repetition rate (~80 MHz) KTP femtosecond OPO. The linear cavity design of the OPO allows the pump light to be reflected into the cavity for double-pass pumping leading to an increase in the output power of the OPO.

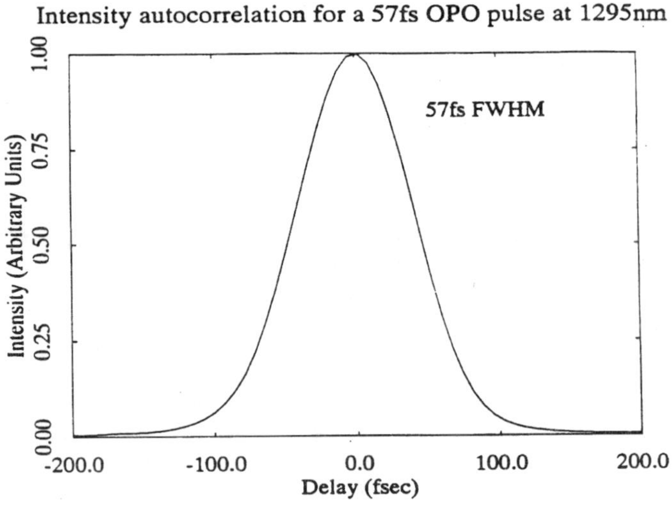


**Fig. 6.9** Intensity autocorrelation trace for a 57 fs Ti:sapphire laser pumped KTP OPO pulse.

Ring and linear cavity configurations of both the pump laser and the OPO cavities have been tried. The relative advantages of the different configurations are as follows: For the pump laser, while the linear cavity is easier to operate, the ring cavity is less sensitive to the light reflected back from the OPO; thus, the two cavities are decoupled without the need for an isolator. In fact, one of the shortest pulses achieved so far in a KTP OPO was with the ring cavity (57 fs, see Figure 6.9 for autocorrelation). For the OPO, the linear cavity has two advantages. It is easier to move the two intracavity prism compensators into and out of the cavity without the need for realigning all the mirrors. Typically, only the output coupler needs to be retouched. Furthermore, with the linear cavity configuration, it is possible to couple back the residual pump beam for a second pass to provide additional pumping. The disadvantage is that for single-pass pumping, the signal pulse sees gain only in one direction but has Fresnel loss and additional group velocity dispersion while traversing the crystal in the unpumped direction. With the ring-cavity for the OPO, the signal pulse sees gain each time it passes through the nonlinear crystal in synchronism with the pump pulse. In all cases, with intracavity prisms to compensate for group velocity dispersion, nearly perfect unchirped transform-limited pulses were obtained.

A major consideration in the design of femtosecond OPO is the inverse pulse group velocity mismatch in the three very different wavelengths (pump, signal, and idler) within the nonlinear crystal. This limits the crystal length to $\sim$1 mm for most inorganic nonlinear crystals. For such a short crystal, the spontaneous parametric emission that can be used for initial alignment purposes is generally quite low, on the order of $10^{-12}$ W in a 1 nm bandwidth. This together with the low single pass gain makes length adjustment and alignment process rather difficult. With practice and experience, it can, however, be done routinely.

The radius of curvature of the focusing mirrors around the nonlinear crystal is determined by the needs to have high enough pump intensity in the crystal and to be able to accommodate the angular acceptance angle to match the entire spectral width of the short pulse within the crystal interaction length. It is on the order of $R \approx 10$ to 15 cm for the OPO-cavity mirrors and 15 to 25 cm for the pump beam focusing mirror.

The tuning characteristics of Ti:sapphire laser pumped KTP OPO is shown in Figure 6.10 for $\phi = 0°$. Tuning of the OPO could be achieved through several means. Certain amount of tuning can be achieved by simply varying the OPO cavity length. Because of group velocity dispersion, the wavelength of the OPO will shift with the cavity length to keep the group velocity and, hence, the cavity round-trip time of the signal pulse constant to match the fixed pump pulse repetition rate. Tuning can of course be achieved by tuning the pump wavelength, as shown in Figure 6.10. One practical advantage of this tuning scheme is that no angular rotation of the crystal or readjustment of the mirrors is required; only length adjustment of the OPO cavity may be needed to match the concomitant change in the cavity round-trip time of the pulse

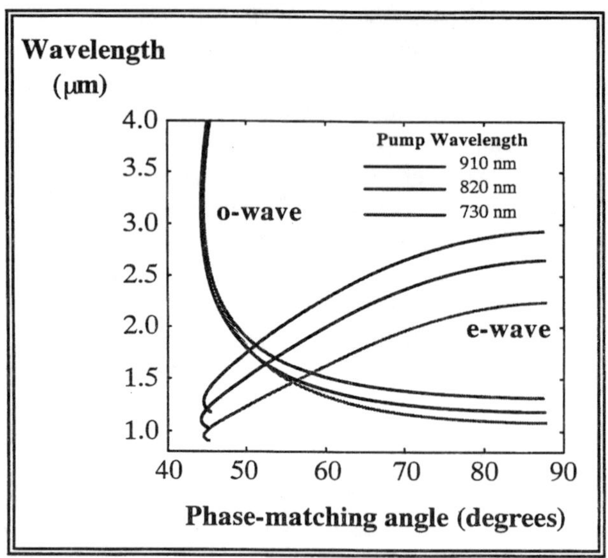


**Fig. 6.10** Tuning characteristics of Ti:sapphire laser pumped KTP femtosecond OPO.

due to group velocity dispersion. This is particularly important for noncritically phase-matched OPO in which the phase-matching condition is satisfied along a principal axis of the nonlinear crystal and, thus, less critical to angular misalignment. For larger tuning range, the phase-matching condition must be altered through changing either the orientation or the temperature of the crystal. Temperature tuning[37] also has the advantage of not requiring any mechanical change in the OPO cavity to maintain noncritical phase-matching, for example.

For femtosecond OPO's, intracavity group velocity dispersion (GVD) of the pump, signal, and idler waves has a strong influence on the temporal and spectral characteristics of the output pulse. It can be adjusted through the intracavity prisms shown in Figure 6.8. With intracavity dye laser pumping, when the OPO is properly adjusted to the near zero group velocity dispersion (GVD) point, the output characteristics are generally quite ideal with a smooth and relatively symmetric spectrum and very clean interferometric autocorrelation trace (Figure 6.7). It shows a close-to-ideal 8:1 ratio indicating good mode-locking, an extremely flat background level indicating pulse amplitude stability, and lack of structures in the wings-indicative of chirp-free operation. With Ti-sapphire laser pumping at a much higher intensity level, the OPO pulses are characterized by a chirped and an unchirped regimes. The chirped regime is encountered when the cavity is operated with a net positive GVD, and the unchirped regime occurs with net negative GVD. With intracavity dispersion compensation it is possible to vary the GVD from net negative values to net positive values. A flip from chirped pulses to unchirped pulses was observed when changed from negative to net positive GVD. At the 0 GVD point the pulses spontaneously jump between chirped and unchirped pulses, leading to instability in the pulse train. In the unchirped regime, highly stable operation of the OPO with ideal interferometric autocorrelation at close to 8:1 ratio can be obtained routinely.

The efficiency and output power of Ti-sapphire laser pumped fs OPO are surprisingly high and the OPO output pulse train is always quite stable for many hours of continuous operation. Pump depletion over 50% due to parametric conversion and total average signal power nearly 700 mW have been observed. An additional remarkable feature of the OPO is that precisely synchronized pulse trains at seven different wavelengths are present simultaneously which include the signal, the idler, the pump, the phase-unmatched second harmonic of the signal and idler, and the phase-unmatched sum of the pump and the signal and the idler. The phase-unmatched second harmonic can be very intense. In fact, almost 100 mW was generated when the total signal power generated was approximately 450 mW. The simultaneous availability of all these wavelengths makes it a particularly useful femtosecond source for doing pump-and-probe experiments.

### 6.2.3 Ti:Sapphire Laser Pumped Femtosecond KTA, CTA, and RTA OPO's

With the demonstration of the KTP fs OPO, it is obvious that many other nonlinear crystals could be used in fs parametric oscillators for other wavelength ranges and to meet other requirements. Fs OPO using three other closely related crystals in the KTP isomorphs $MTiOXO_4$ where $M = \{NH_4, K, Cs$ (for X = As only), $Rb, Tl\}$ and $X = \{P$ or $As\}$ have already been successfully demonstrated. KTA,[38] CTA[39] and RTA[40] are similar to KTP with one important difference. KTP has a rather strong absorption band due to the orthophosphate overtone near 3.5 $\mu$m which prevents the fs OPO to operate efficiently in the important 3–4 $\mu$m range. This overtone absorption band can be eliminated by replacing the phosphate with the heavier orthoarsenate $(AsO_4)^{-3}$ group. KTA, CTA, and RTA can potentially cover the 3–5 $\mu$m spectral range with comparable characteristics as those of the KTP fs OPO. The corresponding tuning curves of these OPO's are shown in Figures 6.11, 6.12, and 6.13.

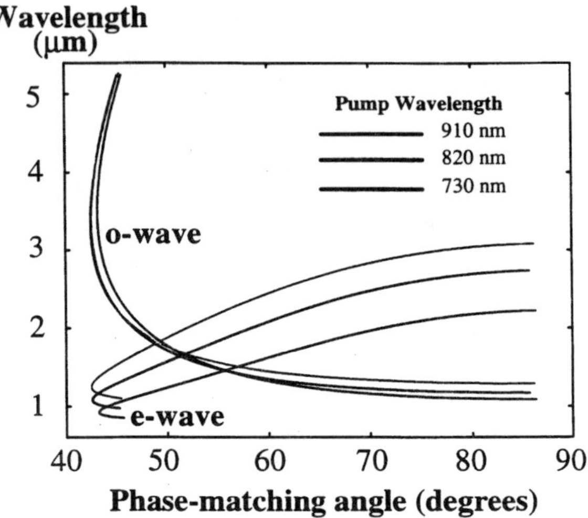


**Fig. 6.11** Tuning characteristics of Ti:sapphire laser pumped KTA femtosecond OPO.

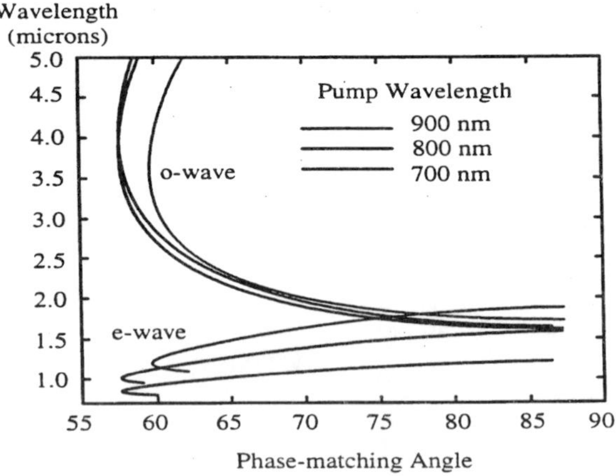


**Fig. 6.12** Tuning characteristics of Ti:sapphire laser pumped CTA femtosecond OPO.

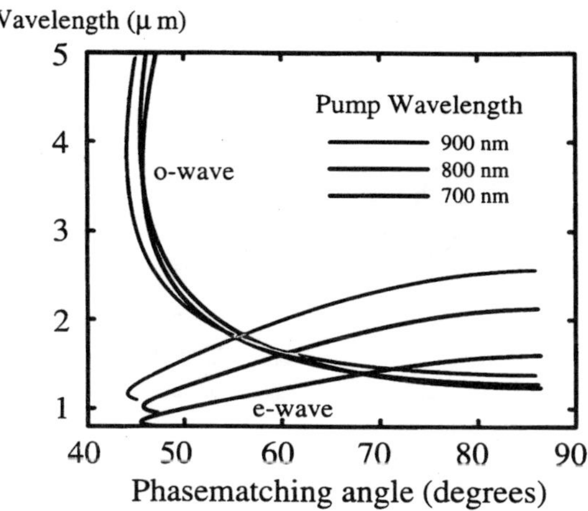


**Fig. 6.13** Tuning characteristics of Ti:sapphire laser pumped RTA femtosecond OPO.

### 6.2.4 Extension into the Mid Infrared, Visible, and Ultraviolet

To go further into the infrared, crystals such as $AgGaS_2$ or $AgGaSe_2$ can be pumped at suitable wavelengths for parametric interaction out to 8 $\mu$m or 18 $\mu$m, respectively. Possible tuning curves are shown in Figures 6.14 and 6.15. The pump requirements in wavelength and in power can be met easily with Ti:sapphire laser pumped OPO's. For femtosecond generation in the infrared, it is also possible to use difference-frequency generation in these crystals using the OPO as the source.

For extension of the output of the fs lasers and the Ti:sapphire laser pumped fs OPO's to the visible and to the ultraviolet, a variety of harmonic or sum-frequency generation processes can be used. One can use intracavity or external cavity doubling or summing with the pump wave of the output of the OPO, or use the doubled or tripled output of a laser as the pump source for the OPO.[41] Some of these have already been demonstrated with impressive results. Particularly successful examples are the intracavity BBO doubled Rh6G dye laser to 315 nm,[42] the Ti:sapphire laser to the blue,[43] and the KTP OPO[44] to cover most of the visible-to-near ir range.

The key issue in frequency-doubling of the femtosecond sources is the need to avoid broadening of the femtosecond pulses at the fundamental wavelength and the second harmonic wavelength. This requires that the doubling crystal be very thin.

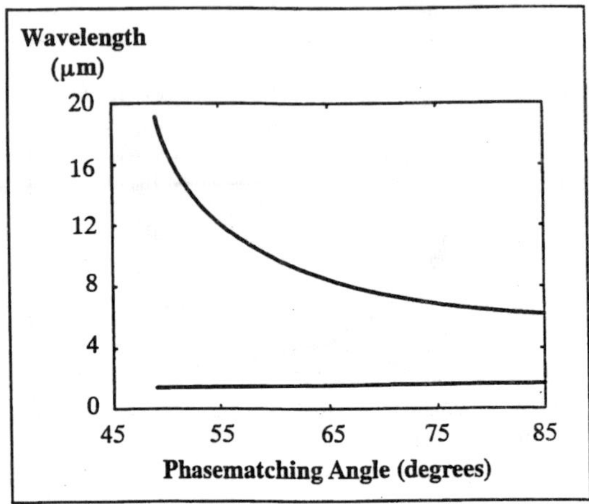


**Fig. 6.14** Theoretical tuning characteristics $AgGaSe_2$ femtosecond OPO pumped at 800 nm with Type II interaction.

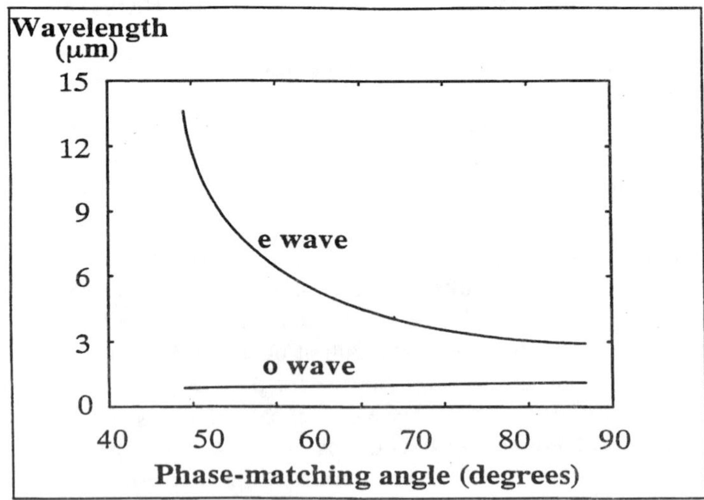


**Fig. 6.15** Theoretical tuning characteristics $AgGaS_2$ femtosecond OPO pumped at 800 nm with Type II interaction.

For very thin crystals, the single-pass conversion efficiency is invariably very low even with very tight focusing, which means that intracavity doubling is often necessary. Because of its uv transparency and phase-matching properties, the crystal of choice for this purpose is $\beta$-barium borate (BBO).

In the experiment on intracavity doubled KTP fs OPO reported in Ref. 44, an intracavity BBO crystal with a thickness of 47 $\mu$m was used. A schematic of such an arrangement is shown in Figure 6.16. An additional focus to accommodate the BBO crystal is introduced into the OPO ring cavity. A ring cavity rather than a linear cavity is used in this case so that the OPO would oscillate in one direction and a single unidirectional second harmonic beam is emitted. Due to the BBO crystals' large second-harmonic generation (SHG) phase-matching bandwidth around the zero group velocity mismatch point at 1.47 $\mu$m, tuning the frequency-doubled OPO output in the range of ~1.1 to ~1.6 $\mu$m (SHG ~550 nm to ~800 nm) requires no adjustment of the BBO phase-matching angle. Continuous tuning of the SHG from 550 to 660 nm limited only by optics available at the time of the experiment has been demonstrated (Figure 6.17 shows data from 580 to 660 nm). Potential tuning range is from ~500 nm to beyond 0.7 $\mu$m where the KTP fs OPO kicks in. The second-harmonic pulse train exhibits excellent stability, and as demonstrated by the real-time interferometric autocorrelation the pulses are chirp-free. Regardless of the transverse mode structure of the Ti:sapphire pump laser, one can achieve an exceptionally clean $TEM_{oo}$ mode for the OPO which is imparted to the intracavity frequency-doubled beam.

High repetition rate OPO operation in the visible down to 13 fs has been reported by G.M. Gale, *et al.*,[45] using the second harmonic of a mode-locked Ti:sapphire laser output as the pump of a BBO OPO.

In conclusion, by using different crystals, pump sources, and frequency-conversion techniques, OPO's will ultimately extend the range of broadly tunable sources of coherent radiation, including high-repetition rate broadly tunable femtosecond pulses, even further into the IR and the UV. Such sources will have a great impact on spectroscopic and other applications.

## 6.3 PICOSECOND OPTICAL PARAMETRIC OSCILLATORS AND AMPLIFIERS

For ultrafast nonlinear optical applications, there is a need for broadly tunable short pulses at peak power or energy/pulse levels much higher than those achievable with current mode-locked high repetition rate continuous-pulse-train laser pumped femtosecond OPO's. These cannot be achieved on a continuous-pulse-train basis at high repetition rates (tens of MHz or higher) basis because of the average-power limitations. To achieve high power in optical parametric devices, there are two choices:

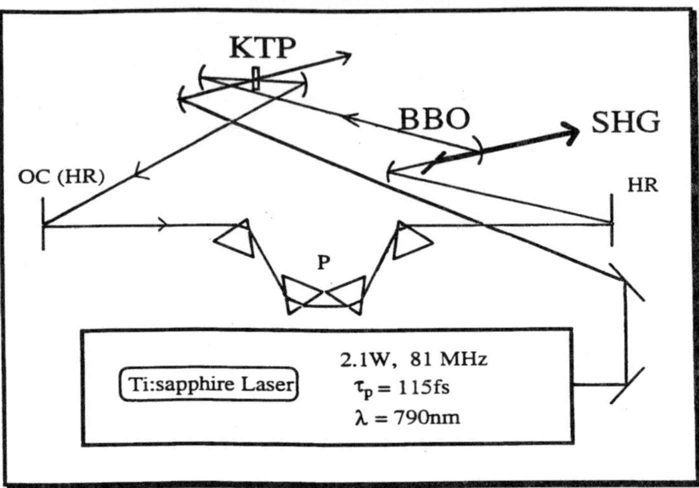


**Fig. 6.16** Schematic of Ti:sapphire laser pumped intracavity doubled KTP OPO. BBO = $\beta$-barium borate doubling crystal. HR = high reflectivity mirrors. P = dispersion-compensation prisms.

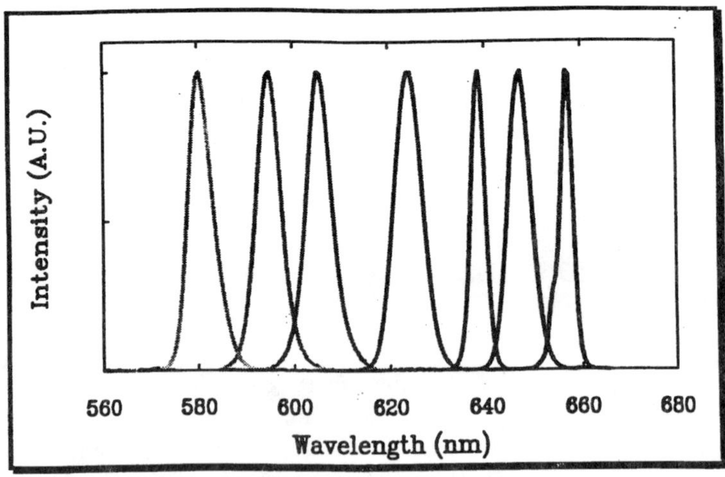


**Fig. 6.17** Spectra showing the demonstrated tuning range of the intracavity frequency-doubled OPO from 580 to 657 nm. (Ref. 43).

1) Use a $Q$-switched mode-locked pulsed solid-state laser such as Nd:YAG or Nd-glass as the pump source for OPO. In this case, the pump and the OPO outputs are of the form of 10–30 Hz bursts of dozens to a few hundred short pulses in the picosecond or femtosecond range.[46–48] 2) Use a low repetition rate (300 KHz or lower) optical parametric amplifier to amplify broad band spontaneous parametric emission.[49,50] The basic limitation in the trade-off between energy per pulse and pulse repetition is that the average power is generally limited to the order of a watt.

With $Q$-switched mode-locked Nd:YAG lasers at pump pulse energy levels of multi-mJ/pulse and repetition rates of 10 to 30 bursts per second, it is possible to reach total gains of 30 to 50 dB through synchronous traveling-wave pumping of several optical parametric amplifiers (OPA) in series. With this kind of gain, it is possible to amplify broadly tunable parametric spontaneous emission up to the mJ/pulse levels for nonlinear optical applications in the mid-picosecond time domain. For pulses shorter than a few psec, it will be difficult to use the mode-locked Nd:YAG laser . For pulse widths shorter than a few psec in the $\mu$J to mJ/pulse range, amplified mode-locked dye laser output pumped femtosecond OPA's operating in the KHz rate range have been demonstrated.

Amplified parametric fluorescence will generally have a relatively broad linewidth without spectral selection. This is less a problem for ultra-short pulse applications where the linewidth requirement is not as stringent as those for high-resolution cw applications. The basic linewidths of the parametric amplification process due to various line broadening mechanisms are given in Section 2.5. In an amplifier, because of the angular dependence of the phase-matching condition, both the parametric emission and the gain are angular dependent. The total linewidth of the amplified parametric emission depends, therefore, on the collection geometry of all the optics along the beam path and must be determined numerically for each specific experimental configuration. Experimentally, with the help of dispersive elements and judicious choice of various relevant parameters, linewidths of the amplified parametric fluorescence over 100 $\mu$J down to the Å level corresponding to near-transform-limited picosecond output of a BBO OPA has been reported.[49]

With the broad range of choices of pump sources (for example, the diode lasers[51]) and nonlinear optical crystals now available, ultrashort femtosecond pulses up to multi-$\mu$J over a large spectral range will surely be available through the parametric processes in the near future. Also there are a variety of interesting theoretical and experimental issues[52–54] directly related to the parametric devices yet to be studied.

The optical parametric amplification process and amplifiers in particular are discussed in detail by Professor Y.R. Shen and co-workers in a companion volume in this series of monographs.

## REFERENCES


1. F.G. Colville, A.J. Henderson, M.J. Padgett, J. Zhang, M.H. Dunn, *Opt. Letters*, **18**, 205 (1993); F.G. Colville, M.J. Padgett, J. Zhang, and M.H. Dunn, *Opt. Letters*, 1065 (1993).
2. D. Lee and N.C. Wong, *J. of Opt. Soc. Am. B*, **10**, 1659 (1993).
3. R.C. Echardt, C.D. Nabors, W.J. Kozlovsky and R.L. Byer, *J. Opt. Soc. of Am. B*, **8**, 646 (1991).
4. H. Vanherzeele, *Appl. Optics*, **29**, 2246 (1990).
5. H.-J. Krause and W. Daum, *Appl. Phys. Lett.*, **60**, 2180 (1992).
6. S. Lin, J.Y. Huang, J. Ling, C. Chen and Y.R. Shen, *Appl. Phys. Lett.*, **59**, 2805 (1991).
7. J. Chung and A.E. Siegman, *J. Opt. Soc. Am. B*, **10**, 2201 (1993).
8. A. Nebel, C. Fallnich, R. Beigang and R. Wallenstein, *J. Opt. Soc. of Am. B*, **10**, 2195 (1992).
9. G.A. Rines, D.M. Rines and P.F. Moulton, *OSA Conference on Advance Solid-State Lasers*, Technical Digest 280 (1994).
10. S. Chandra, M.J. Ferry and G. Daunt, *OSA Proceedings on Advanced Solid-State Lasers*, **13**, 353 (1992).
11. A. Fix, T. Schröder, R. Wallenstein, B.J. Orr, M.J. Johnson and J.G. Haub, *J. Opt. Soc. Am. B*, **10**, 1744 (1993).
12. A.J. Campillo and C.L. Tang, *Appl. Phys. Lett.*, **19**, 36 (1971).
13. R.L. Byer in *Book Optical Parametric Oscillator*, H.T. Rabin and C.L. Tang, Eds., Vol. B, p. 588, Academic Press: (1975).
14. Y.X. Fan, R.C. Eckardt, R.L. Byer, C. Chen, and A. Jiang, post deadline paper ThT4, CLEO '86, Anaheim, CA; Y.X. Fan, R.C. Eckardt, R.L. Byer, J. Nolting, and R. Wallenstein, *App. Phys. Lett.*, **53**, 2014 (1988).
15. L.K. Cheng, W.R. Bosenberg, and C.L. Tang, *App. Phys. Lett.*, **53**, 175 (1988); *App. Phys. Lett.*, **54**, 13 (1989); L.K. Cheng, "*Growth and application of low temperature phase β-barium borate crystals*", Cornell University Ph.D. thesis, 1988.
16. W.R. Bosenberg, W.S. Pelouch, and C.L. Tang, *App. Phys. Lett.*, **55**, 1952 (1989); W.R. Bosenberg, "*Development of the β-barium borate optical parametric oscillator*", Cornell University Ph.D. thesis, 1990.
17. T.K. Minton, S.A. Reid, H.L. Kim and J.D. McDonald, *Opt. Commun.*, **69**, 289 (1988).
18. W.R. Bosenberg and D.R. Guyer, *Appl. Phys. Lett.*, **61**, 387 (1992).
19. W.R. Bosenberg and C.L. Tang, *App. Phys. Lett.*, **56**, 1819 (1990).
20. X. Liu, D. Deng, M. Li, D. Guo and Z. Xu, *J. Appl. Phys.*, **74**, 2989 (1993).
21. M. Ebrahimzadeh, G.J. Hall and A.I. Ferguson, *Opt. Lett.*, **17**, 652 (1992).
22. J.G. Haub, M.J. Johnson, and B.J. Orr, *App. Phys. Lett.*, **58**, 1718 (1991).
23. W.R. Bosenberg and D.R. Guyer, *J. Opt. Soc. Am. B*, **10**, 1716 (1993).
24. H. Komine, *J. Opt. Soc. Am. B*, **10**, 1751 (1993).
25. See Reference 135–137 in Chapter 5.
26. P.G. Schunemann, P.A. Budni, M.G. Knights, T.M. Pollak, E.P. Chicklis and C.L. Marquardt, in *Topical Meeting on Advanced Solid State Lasers*, New Orleans, LA., 1993 (Optical Society of America), *2*, 131.
27. W.R. Bosenberg, L.K. Cheng and J.D. Bierlein, in *Topical Meeting on Advanced Solid State Lasers*, New Orleans, 1993 (Optical Society of America), **2**, 134.
28. R.C. Eckardt, X.Y. Fan, R.L. Byer, C.L. Marquardt, M.E. Storm and L. Esterowitz, *Appl. Phys. Lett.*, **49**, 608 (1986).
29. N.P. Barn, K.E. Murray and G.H. Watson, *Proceedings on Advanced Solid-State Lasers*, **13**, 356 (1992).
30. D.C. Edelstein, E.S. Wachman, C.L. Tang, *App. Phys. Lett.*, **54**, 1728 (1989).

31. W.S. Wachman, D.C. Edelstein, and C.L. Tang, *J. App. Phys.*, **70**, 1893 (1991); *Opt. Lett.*, **15**, 136 (1990).
32. D.E. Spence, P.N. Kean, and W. Sibbett, *Opt. Lett.*, **16**, 42 (1991)
33. D.C. Edelstein, *New sources and techniques for ultrafast lasers spectroscopy*, Cornell University, Ph.D. Thesis, (1990); E.S. Wachman, "Ultrafast spectroscopy with a novel broadly tunable cw femtosecond source", Cornell University, Ph.D. thesis, 1991.
34. L.A.W. Gloster, Z.X. Jiang, and T.A. King, *J. Quant. Elect.*, (to be published).
35. W.S. Pelouch, P.E. Powers, and C.L. Tang, *Opt. Lett.*, **17**, 1070 (1992),
36. Q. Fu, G. Mak, and H. Van Driel, *Opt. Lett.*, **17**, 1006 (1992).
37. J.D. Kafka, M.L. Watts, and J.W. Perterse, post deadline paper, CLEO, Baltimore, 1993; S.D. Butterworth, M.J. McCarthy, and D.C. Hanna, (to be published *Opt. Lett.*, Sept. 93).
38. P.E. Powers, S. Ramakrishna, C.L. Tang, and L.K. Cheng, *Opt. Lett.*, **18**, 1171 (1993).
39. P.E. Powers, C.L. Tang, and L.K. Cheng, *Opt. Lett.*, **19**, 37 (1994).
40. P.E. Powers, C.L. Tang, and L.K. Cheng, *Opt. Lett.* (to be published, 1994).
41. T.J. Driscoll, G.M. Gale, and F. Hache, *Opt. Comm.* (to be published).
42. D.C. Edelstein, E.S. Wachman, L.K. Cheng, W.R. Bosenberg, and C.L. Tang, *App. Phys. Lett.*, **52**, 2211 (1988).
43. R.J. Ellingson and C.L. Tang, *Opt. Lett.*, **17**, 343 (1992).
44. R.J. Ellingson and C.L. Tang, *Opt. Lett.*, **18**, 438 (1993).
45. G. Gale, M. Cavallari, T. Driscoll and F. Hache, *Opt. Lett.*, **20**, 1562 (1995).
46. R. Laenen, H. Graener, and A. Laubereau, *Opt. Lett.*, **15**, 971 (1990) and references to earlier work in W. Kaiser's group at Munich.
47. V. Kubecek, Y. Takagi, K. Yoshihara and G.C. Reali, *Opt. Commun.*, **91**, 93 (1992).
48. W.T. Lotsahaw, J.R. Unternahrer, M.J. Kukla, C.I. Miyake and F.D. Braun, *J. Opt. Soc. of Am. B*, **10**, 2191 (1993).
49. J.Y. Huang, J.Y. Zhang, Y.R. Shen, C. Chen, and B. Wu, *App. Phys. Lett.*, **57**, 1961 (1990).
50. R. Danielius, A. Piskcarskas, A. Stabinis, G.P. Banfi, P.D. Trapani, and R. Righini, *J. Opt. Soc. Am. B*, **10**, 2222 (1993); F. Seifert, V. Petrov, and Noack, *Opt. Lett.*, **19**, 837 (1994).
51. M.J. McCarthy and D.C. Hanna, *J. of Opt. Soc. Am. B* (Nov., 1993); M.J. McCarthy, S.D. Butterworth, and D.C. Hanna (to be published in Opt. Comm.).
52. E.C. Cheung and J.M. Liu, *J. of Opt. Soc. of Am. B*, **8**, 1491 (1991).
53. D.G. Reid, J.M. Dudley, M.N. Ebrahimzadeh, and W. Sibbett, *Opt. Lett.*, **19**, 825 (1994).
54. J.D.V. Khaydarov, J.H. Andrews, and K.D. Singer, *Optics Lett.*, **19**, 831 (1994).

# Index


Average power barrier, 101

Biaxial crystals, 13, 62–67
Birefirngence, 10, 11
Bloembergen symmetry condition, 58

Conservation of
  Energy, 4, 9
  Momentum, 5, 9
Coupled-amplitude equations, 17
Coupled-mode equations, 16, 19
Coupled-wave equations, 17

Dispersion, 10, 11

Effective Kleinman d-coefficients, 18
  BBO, 59
  Biaxial crystals, 62–65
  KTP, 76
  $LiB_3O_5$, 92, 93
  Parametric process, 58, 59
  Second-harmonic generation, 58, 61, 72, 83, 86
Equivalent noise temperature, 2, 22

Femtosecond OPO, see *Optical Parametric Oscillators*

Intracavity doubling, 123, 124

Manley-Rowe relations, 24
Measurement of $\chi^{(2)}$, 29, 59

Nonlinear polarization, 3, 18

Optically pumped three-level lasers, 5, 6
OPO materials properties and characterization, 53
  ADP, 14
  $AgGaS_2$, 53, 58, 68, 69, 71
  $AgGaSe_2$, 55, 68, 69
  $\beta$-$BaB_2O_4$, 85–90
    Phase and group velocities, 55
  $CdGaAs_2$, 68
  Chalcopyrites, 68
  GaSe, 68, 71
  KDP, 72
  $KNbO_3$, 56, 68, 70, 74, 81–85
  $KTiOPO_4$ and its isomorphs, 54–57, 68, 70, 73–81
  *L*-arginine phosphate, 53
  $LiIO_3$, 70
  $LiB_3O_5$, 53, 87, 90–94
  MgO:$LiNbO_3$, 53, 56, 57, 67, 70, 72, 74
  $Tl_3AsSe_3$, 55, 58, 69–72
  $ZnGeP_2$, 53, 54, 68, 69, 71
  Urea, 56, 57, 68, 87
Optical parametric oscillators (OPO), 37
  Back conversion, 46
  Clustering effect in DRO, 42
  Femtosecond OPO, 56, 106–123
    KTA, CTA, RTA OPO, 119–123
    Rh6G dye laser pumped KTP OPO, 108–115
    Ti:sapphire laser pumped KTP OPO, 115–119
  Intracavity intensity, 44, 45,
  $KTiOPO_4$ and its isomorphs, 105–121
  Large-signal theory, 42
  $LiB_3O_5$, 105
  Linewidth, 49–51, 104, 105
  Nanosecond OPO, 56, 102–106
    $\beta$-$BaB_2O_4$, 103–106
  Oscillation threshold conditions, 38
    Singly-resonant (SRO), 38
    Doubly-resonant (DRO), 40
  Quantum efficiency, 44, 46
  Schematics of
    BBO OPO, 39
    KTP fs OPO, 109, 116

Synchronous pumping, 37, 108
Threshold condition, 46, 47
Walk-off effect, 40, 66, 103, 111–113

Parametric amplification
Back-conversion, 25
Bandwidth due to
Finite crystal length, 26
Pump bandwidth, 26
Pump beam divergence, 25
Classical theory, 15
Large signal theory, 23
Viewed as repeated difference-frequency generation process, 15
Parametric luminescence or fluorescence, 21
Semiclassical theory, 22
Phase matching condition, 9
Biaxial crystals, 64
Noncritical, 13
Quasi-phase-matching, 15, 26
Uniaxial crystals
Type I, 10, 12
Type II, 12, 13
Photon flux density, 20
Picosecond OPO's and OPA's, 123–125

Quantum efficiency of OPO, 7, 44, 46
Quantum theory of spontantous parametric scattering, 29-35
Field Hamiltonian, 30, 31
Initial and final states, 31
Interaction Hamiltonian, 30, 31
Rules of quantization and commutation relationships, 30, 31
Transition probability, 31

Repeated difference-frequency generation process, 9
Model for parametric amplication process, 15

Saturation effects, 16, 42–49
Signal and idler waves, 9, 10
Slowly-varying-amplitude approximation, 19
Spatial gain coefficient, 16, 20
Spontaneous scattering processes
Brillouin, 21
Parametric, 21, see also *parametric luminescence or fluorescence*
Polariton, 21
Raman, 21
Squeezing, 2, 29, 36

Three-photon parametric process, 4
Feynman diagram, 3
Final state, 4
Initial state, 4
Spontaneous emission, 4, 21, 33, see also *Spontaneous parametric emission or scattering Process, or Parametric luminescence or fluorescence*
Stimulated emission, 4
Tunability, origin of, 4
Tuning characteristics, examples of
$AgGaS_2$, 122
$AgGaSe_2$, 122
BBO OPO vs. common lasers, 2
BBO OPO, 12
$KNbO_3$, 84
$KTiOPO_4$ and its isomorphs, 81, 110, 118, 120, 121
$LiB_3O_5$, 92, 93
Noncollinear, 14
$Tl_3AsSe_3$, 72
Type II of CTA and KTA, 81
Tuning range of OPO, 2
Tuning schemes
Angular tuning, 13
Temperature tuning, 13

Uniaxial crystals
Negative, 10
Positive, 13

Zero-point fluctuations, 21

# LASER SCIENCE AND TECHNOLOGY
## An International Handbook

SECTIONS
Chaos and Laser Instabilities
Coherent Sources for VUV and Soft X-Ray Radiation
Distributed Feedback Lasers
Excimer Lasers
Fiber Optics Devices
Frequency Stable Lasers and Applications
Gas Lasers
Interaction of Laser Light with Matter
Laser Diagnostics in Chemistry
Laser Fusion
Laser Monitoring of the Atmosphere
Laser Photochemistry
Laser Spectral Analysis
Lasers and Communication
Lasers and Fundamental Physics
Lasers and Nuclear Physics
Lasers and Surfaces
Lasers in Medicine and Biology
Mechanical Action of Laser Light
New Solid State Lasers
Optical Bistability
Optical Computers
Optical Storage and Memory
Phase Conjugation
Semiconductor Diode Lasers
Solid State Lasers
Squeezed States of Light
Topics in Nonlinear Optics
Topics in Theoretical Quantum Optics
Tunable Lasers for Spectroscopy
Ultrashort Pulses and Applications
Vapour Deposition of Materials on Surfaces

# PUBLISHED TITLES

Volume 1 (Lasers in Medicine and Biology Section)
**LASER MICROIRRADIATION OF CELLS**
by T. Kasuya and M. Tsukakoshi

Volume 2 (Lasers and Surfaces Section)
**NONLINEAR OPTICAL DIAGNOSTICS OF LASER-EXCITED SEMICONDUCTOR SURFACES**
by S.A. Akhmanov, N.I. Koroteev and I.L. Shumay

Volume 3 (Ultrashort Pulses and Applications Section)
**LIGHT PULSE COMPRESSION**
by W. Rudolph and B. Wilhelmi

Volume 4 (Laser Spectral Analysis Section)
**ANALYTICAL ASPECTS OF ATOMIC LASER SPECTROCHEMISTRY**
by K. Niemax

Volume 5 (Lasers and Nuclear Physics Section)
**INVESTIGATION OF SHORT-LIVED ISOTOPES BY LASER SPECTROSCOPY**
by E.W. Otten

Volume 6 (Fiber Optics Devices Section)
**NONLINEAR EFFECTS IN OPTICAL FIBERS**
by E.M. Dianov, P.V. Mamyshev, A.M. Prokhorov and V.N. Serkin

Volume 7 (Distributed Feedback Lasers Section)
**OPTICAL OSCILLATORS WITH DEGENERATE FOUR-WAVE MIXING (DYNAMIC GRATING LASERS)**
by S. Odoulov, M. Soskin and A. Khizhniak

Volume 8 (Lasers in Medicine and Biology Section)
**PHOTOBIOLOGY OF LOW POWER LASER THERAPY**
by T. Karu

Volume 9 (Topics in Nonlinear Optics Section)
**REFRACTIVE NONLINEARITY OF WIDE-BAND SEMICONDUCTORS AND APPLICATIONS**
by A.A. Borshch, M. Brodin and V. Volkov

Volume 10 (Laser Fusion Section)
**INTRODUCTION TO LASER FUSION**
by C. Yamanaka

Volume 11 (Lasers and Fundamental Physics Section)
**PHOTOCHEMICAL LASERS**
by V.S. Zuev and L.D. Mikheev

Volume 12 (Lasers and Fundamental Physics Section)
**LASERS IN ACOUSTICS**
by F.V. Bunkin, A.A. Kolomensky and V.G. Mikhalevich

Volume 13 (Interaction of Laser Light with Matter Section)
**INTERACTION OF INTENSE LASER LIGHT WITH FREE ELECTRONS**
by M.V. Federov

Volume 14 (Distributed Feedback Lasers Section)
**THEORIES ON DISTRIBUTED FEEDBACK LASERS**
by F.K. Kneubühl

Volume 15 (Topics in Nonlinear Optics Section)
**DEVELOPMENT OF NEW NONLINEAR OPTICAL CRYSTALS IN THE BORATE SERIES**
by C.T. Chen

Volume 16 (Tunable Lasers for Spectroscopy Section)
**ROOM TEMPERATURE TUNABLE COLOR CENTER LASERS**
by T.T. Basiev and S.B. Mirov

Volume 17 (Laser Fusion Section)
**LASER PLASMA THEORY AND SIMULATION**
by H.A. Boldis, K. Mima, A. Nishiguchi, H. Takabe and C. Yamanaka

Volume 18 (Mechanical Action of Laser Light Section)
**ATOM OPTICS WITH LASER LIGHT**
by V.I. Balykin and V.S. Letokhov

Volume 19 (Topics in Nonlinear Optics Section)
**OPTICAL PARAMETRIC GENERATION AND AMPLIFICATION**
by J. Zhang, J.Y. Huang and Y.R. Shen

Volume 20 (Topics in Nonlinear Optics Section)
**FUNDAMENTALS OF OPTICAL PARAMETRIC PROCESSES AND OSCILLATORS**
by C.L. Tang and L.K. Cheng

Further volumes in preparation.